CIVIL ENGINEERING
PRACTICE

CIVIL ENGINEERING PRACTICE
An introduction

Dr Stephen Scott
University of Newcastle upon Tyne, UK

A member of the Hodder Headline Group
LONDON • SYDNEY • AUCKLAND

First published in Great Britain 1997 by
Arnold, a member of the Hodder Headline Group,
338 Euston Road, London, NW1 3BH

British Library Cataloguing in Publication Data
A catalogue record for this book is available from the British Library

ISBN: 0 340 69273 1

Typeset in 10/12pt Times by
Anneset, Weston-super-Mare, Somerset
Printed and bound in Great Britain by
J W Arrowsmith Ltd, Bristol.

Contents

Preface

The memories of the complexity of the course I took as a civil engineering student back in the late 1960s and of the considerable demands it made on me are still clear. I confess that most of my memories of that time now fall into this general category. After several years of not using the information, little of the specifics of subjects such as structural engineering and even less of fluid mechanics remain in accessible storage. Nevertheless, the requirement that civil engineers should be trained by exposing them to all of the disciplines of the profession is difficult to fault, and yet with the increased volume of knowledge academics put into their courses, the burden on the student can only have increased.

Another strong memory from those days was the constant worry that, according to most of my lecturers, I was meant to be 'reading' for my degree and yet it seemed quite difficult enough to be reading the copious notes I was making in lectures and trying to make some kind of sense of them. Visits to the library at that time seldom seemed very fruitful, with hours being spent to make little progress. From my current position as an academic in a civil engineering department, it is clear that these problems still exist. Ideally, all students should be reading around their subject, but in practice, given the strain of most engineering courses, this is unlikely to happen for the majority. There will, of course, be occasions where course-work or project work properly require that the essential skills of searching for information be learnt but the number of occasions when students delve into books other than to understand better the material they have received in lectures is likely to be small. The thirst for knowledge is clearly not a great motivator when you haven't got time to drink.

It was with this view of the lot of most engineering students that this book was written. The intention was to cover a useful body of work which would give the reader a good introduction to the typical practices followed in the majority of civil engineering projects, without attempting to say everything that could be said on each subject. The material in the book developed over a number of years as a lecture course for students, mainly undergraduates, and I believe has benefited from that process. By taking note of the particular aspects of each topic that regularly confused the students on the course and making special efforts to explain those areas carefully, the final text will hopefully be more accessible as a result.

1 Civil engineering procedure

Introduction

Most civil engineering involves the design and construction of projects and brings together a number of separate parties to achieve this. In this first chapter, an overview is given of the parties involved and the stages through which a project typically progresses from inception to final completion. These stages, together often known as the project cycle, are depicted in Fig. 1.1. It should be realised that this diagram is a very simplified view of the procedure and represents what is often called a 'traditional' arrangement of the parties. This involves a Promoter who employs an Engineering team to design the works and to supervise a Contractor, who constructs the works. Any arrangement of the parties that does not follow this traditional layout would produce a diagram different from Fig. 1.1, but as this arrangement is quite common, it makes a useful framework around which to discuss the functions and responsibilities of the parties involved.

Parties to the project

The Promoter

Most civil engineering projects involve the construction, redevelopment or maintenance of parts of the infrastructure and these are usually termed 'public-sector' projects. In these, the Promoter will typically be a government department (e.g. the Department of Transport) or a local authority. Promoters in the private sector are more likely to be developing their own assets, often in the form of office buildings, factories or warehouses. It is always the Promoter who wants the project and who must have sufficient finances to be able to procure the project.

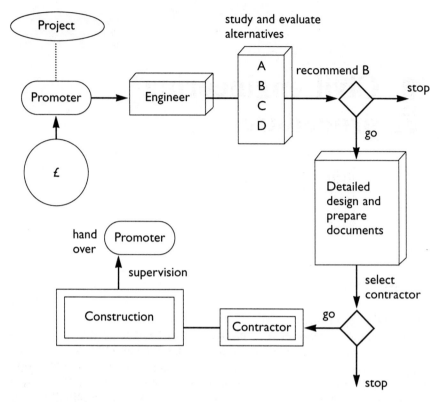

Fig. 1.1 Realisation of a civil engineering project

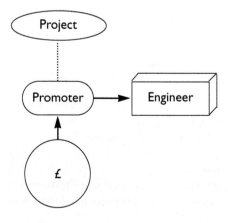

Fig. 1.1a

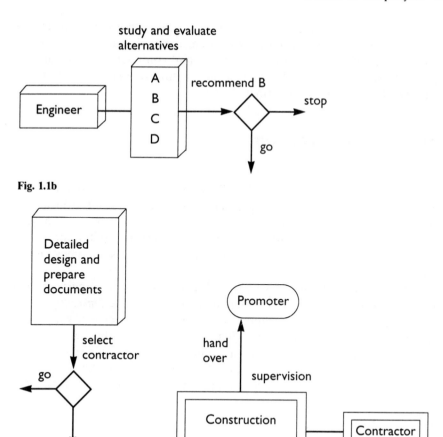

Fig. 1.1b

Fig. 1.1c   **Fig. 1.1d**

The Engineer

The term 'Engineer' is often used to describe the organisation that designs the project and may also supervise its construction. By supervision is meant the inspecting and checking of the works that will be carried out by the contractor. Some Promoters have their own in-house staff who can fulfil this function, but the recent trend is towards the use of a private-sector firm of consulting engineers. Note that such a firm will often employ many *engineers* amongst its staff.

The Contractor

Construction of the project will be carried out by a separate organisation specialising in the mobilisation and management of men and resources to bring about the successful completion of the work. Most such organisations are in the private sector, although there are still some direct labour organisations (DLOs), mostly working for local authorities, who do this kind of work. Complex projects are often nominally carried out by one contractor, but it is common for these 'main' contractors to subcontract parts of the work to specialist subcontractors; for example, piling work and earthworks are often subcontracted.

Stages in the procedure

The Promoter, who is aware of the need for a project and can finance it, but does not have the technical skill to develop it further, employs the Engineer to help progress the project. At this early stage, the Promoter must define the essential features of the project as completely as possible so that the Engineer's efforts are properly focused. This definition of the project and the extent of the Engineer's involvement is sometimes known as the *design brief*.

With an understanding of what the Promoter wants, the Engineer then proceeds to the first stage of design, often known as the *feasibility study*. This involves identifying a number of alternative schemes that will satisfy the Promoter's requirements, defining the schemes in outline (*see* Fig. 1.2) and comparing the schemes on a number of factors, viz. cost, public response, environmental impact, etc., to select the best alternative. The Engineer then writes a report detailing the study carried out and the

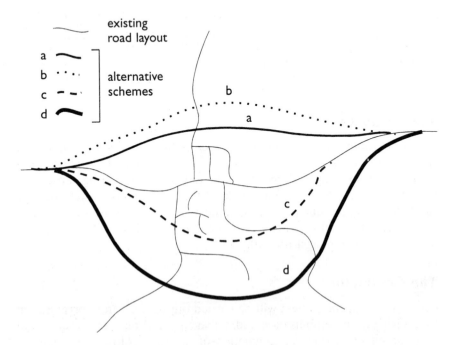

Fig. 1.2 Alternative solutions for a bypass project

scheme recommended. At this stage, the Promoter may stop the project or, more likely, proceed to the next stage.

Now the selected scheme must be designed in detail, with full calculations to fix the exact location of the project, the dimensions of its individual parts and the materials and level of workmanship that will be needed. The completion of the design process culminates in the production of a number of documents (*see the section* 'Tender/contract documents'), and it is on the basis of these documents that a contractor will be selected to carry out the work (*see* 'Arranging a contract'). Having received a firm offer to construct the project, the Promoter may even at this stage decide to go no further. More likely, however, he/she will proceed to the construction stage.

Having entered into a contract with a contractor, the Promoter now relies on the Engineer to supervise the Contractor's work to ensure that it is done in accordance with the contract documents. This part of the procedure is quite complex and will be dealt with in more detail in Chapter 7, 'Conditions of contract'. When the work is complete, it is handed over to the Promoter and the procedure is essentially concluded.

Identification of alternatives

Let us suppose that the project is to alleviate traffic congestion in a town, and the alternative schemes considered as possible solutions to the problem are different routes for a bypass, a, b, c and d (as Fig. 1.2). At the feasibility study stage, these would be identified in outline only, and the initial studies that would be carried out would include land surveys, assessment of ground conditions and the recording of relevant physical data on each of the separate routes. The depth of these investigations need only be sufficient to compare the alternatives; for example, at this stage, ground conditions may be assessed mainly by referring to geological maps. A much more thorough study will of course be undertaken of the preferred route when that is actually selected.

Tender/contract documents

The tender/contract documents are intended to define fully what is to be constructed, how it will be paid for and what risks and responsibilities the various parties must bear. Subsequent chapters will provide more detail about individual documents, but for the moment a short overview will suffice. The documents that must be produced by the Engineer will depend to some extent on the type of contract that is used (*see later*), but for a bill of quantities contract, the following documents will be needed:

Document	Description
Form of tender	The Contractor's *offer* to construct the project
Conditions of contract	The *rules* that will govern the construction of the project
Specification	The *details* of materials and workmanship
Bill of quantities	The way in which *payment* will be made to the Contractor
Drawings	The *job* itself; layout and details of the various parts of the project
Form of agreement	The *deal* signed by the Contractor and the Promoter

It is also common to bundle with these documents another document, commonly called 'instructions to tenderers', which explains to contractors what they must do to submit an acceptable tender. The Instructions are just a guide for use in the tendering process and will not be considered further once tenders have been received.

Arranging a contract

The Promoter wants to have his/her project constructed by a contractor who is competent, financially secure and offering to do the work at a reasonable price. To achieve this last requirement, it is essential that there is some element of competition. The Engineer will identify a number of contractors who are considered to be capable of doing the work and usually four to six of these contractors will be invited to submit tenders (Fig. 1.3). (A tender is a contractor's offer to carry out the works, which will certainly include the price to be paid if the offer is accepted.) Each of the contractors selected will be sent a set of documents, and will be given a period of time in which to study the documents and at the end of the period submit their tenders. These are then assessed by the Engineer, who will typically consider the three lowest tenders and recommend to the Promoter which should be accepted. Provided the project is to go ahead, the Promoter then enters into a contract with the selected contractor and construction can proceed. Clearly only one of the contractors can be successful, and thus the contractors who have been unsuccessful will have to be informed that their tenders will not be accepted: they have failed to win the contract.

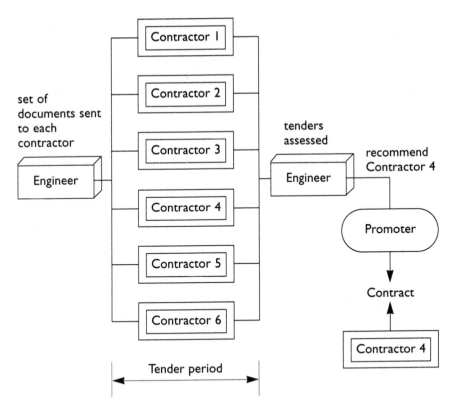

Fig. 1.3 Selecting a contractor

Recommendations for further reading

Institution of Civil Engineers (1986) *Civil Engineering Procedure*, 4th edition.
  Thomas Telford, London. 112 pp.
  A more detailed overview of the typical procedure for realising a civil engineer-
ing project.

2 Types of contract

Introduction

As already mentioned, it is the Promoter's funds that are being used to finance the project and invariably these are limited, so it is important that they be used to best effect. One of the decisions that will affect the cost of the project and that must be made very early in the project's life is the type of contract that will be used, and it is normally expected that the Promoter's advisers will help in making this decision. There will always be some dispute as to which is the best type of contract for a project, and indeed, there seems to be some element of fashion about what is considered the best type to use at any particular time. There are, however, good reasons why certain projects should be dealt with in particular ways, and the alternative general approaches are what will be discussed under this heading.

Some of the types of contract are distinguished by the way in which the Contractor is paid, while others represent a different way of arranging the parties who contribute their funds or expertise to the project. The main types of contract usually recognised are as follows:

Different ways of paying the Contractor

(i)   Lump sum contract
(ii)   Measurement contracts: (a) Bill of quantities
                                             (b) Schedule of rates

(iii)   Cost-reimbursement contracts: (a) Cost + fixed fee
                                             (b) Cost + percentage fee
                                             (c) Target cost

Different ways of arranging the parties

(iv)   Management contract

(v)   Design and build contract

(vi) 'BOOT' contract

The use of BOOT contracts in the UK is a fairly recent innovation and their categorisation is not well defined. They are certainly a different way of arranging the parties, but it is difficult to define a standard arrangement, as contracts let to date vary considerably.

Note that the second class of contracts, above, will usually employ one or more of the first type; for example, design and build contracts will often use a lump sum approach for paying the Contractor, Management contracts may adopt lump sum, bill of quantities and cost-reimbursement methods to make payments.

Different ways of paying the Contractor

(i) Lump sum contract

In the case of a lump sum contract, the Contractor agrees to carry out the work of the project for a sum of money, which will be paid either in full when the project is finished or in instalments when defined stages of the work are completed. The size of the lump sum tendered is clearly an important factor when selecting the Contractor for the project.

For contracts operated on a traditional basis, the Engineer will have designed the works, which will be defined by drawings and a specification. Each Contractor who tenders must understand the work involved in the job by taking off quantities from the drawings and submitting a price. There is thus a considerable duplication of effort with all tenderers having to go through this procedure, and this does not occur with most other types of contract. However, provided the original requirements are not changed, no detailed accounting or measuring is necessary during the course of the contract as the Contractor's payment is already decided. There are less likely to be changes when:

1. the project is not too large;

2. the work required can be precisely described.

Although point 1 above states that this approach is best suited to smaller contracts, this is only so when a traditional arrangement of the parties is being used. For other methods – for example, design and build – a lump sum contract will often be used, and this may be adopted on some of the very largest projects. In these cases, no detailed design exists when the Contractor is being asked to give a price (which is for both the design and the construction of the project), and asking for a lump sum tender is one of only two sensible approaches.

(ii) Measurement contracts

(a) Bill of quantities

The detailed design of the project having been completed, one of the documents that must then be prepared by the Engineer, when using this method, is the bill of quantities. The work of the contract is broken down into categories, e.g. earthworks, drainage, in-situ concrete, and within each category items are provided that cover the job content. The Engineer inserts against each of these items an approximate quantity by measuring the work from the drawings. Figure 2.1 is an example of part of a page of a bill of quantities. When all of the work of the contract has been dealt with in this way, the result is the work items section of the bill of quantities. The way in which the contract work is split into items is defined in a method of measurement (*see* the discussion of CESMM3 in Chapter 4), and in order to submit an acceptable tender, contractors must insert rates against each of these items, calculate the total cost for each item, sum the item costs on each page and collect the totals from each page. It is on this basis that tenders will be compared, and the contractors can expect to be able to rely on the Engineer's bill of quantities as properly representing the real work of the project. Notice that the measurement of the work of the job is done only once, by the Engineer, unlike in the lump sum approach.

A bill of quantities is usually used with a traditional arrangement of the parties and is called a measurement contract because the work of the

Class F: in-situ concrete

No.	Item description	Unit	Quantity	Rate	£	p
	Provision of concrete					
F263	Designed mix grade C30, cement to BS 12, 20 mm aggregate	m^3	48	35.50	1704	00
F293	Designed mix grade 7, blinding concrete	m^3	4	32.00	128	00
	Placing of concrete					
F624	Mass blinding, thickness 75 mm	m^3	4	12.40	49	60

Fig. 2.1 Page from a bill of quantities

contract is remeasured as construction proceeds. Thus the estimates produced by the Engineer in the tender bill of quantities will all be reassessed. Each month, the work carried out is remeasured and the bulk of the payments to the Contractor are made by inserting the quantities of work actually carried out at that date in the bill of quantities and calculating the amounts payable. At the end of the contract, all items in the bill will have been recalculated to give the final quantities against each.

Another advantage of the bill of quantities can be seen when changes have to be made to the original requirements that the Contractor priced. Where work has been added, payment for the additional work must be made to the Contractor and where the extra work is the same as work covered by items in the original bill, the Engineer can expect to pay at the bill rate.

It should be realised by now that a bill of quantities can only sensibly be used where the detailed design of the project is either complete or substantially complete. The main reason for this is that without the design being fully developed, the quantities of work in the job will not be accurately known and the Engineer's estimates of quantities will not be reliable. To price an element of work accurately, the Contractor needs to have a good understanding of the quantity of that particular element in the job. It might be thought that the rate for doing work is not affected by the quantity of work, but for some activities this will not be true. Figure 2.2 shows the way in which the rate for work may vary with the quantity to be carried out; here, because there is a set charge for doing the work irrespective of how much is actually done (say, for hire of a piece of plant

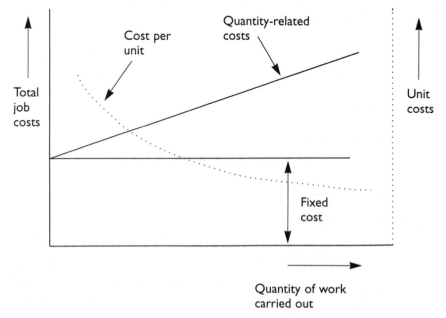

Fig. 2.2 Effect of quantity on unit cost

which can only be hired for a minimum period of one week). Also, learn-ing curve effects will reduce the cost per unit where more units are to be processed. This means that a bill of quantities would never be a sensible option for a project being carried out under a design and build contract.

(b) Schedule of rates

In this approach, instead of a bill of quantities, a list of items of work will be provided by the Engineer for the Contractor to price, but either there will be no quantities quoted or the quantities will be 'ball-park' figures. This will be because the work of the project is not fully defined at the time of tender, something that may stem from a wish to start work early, before the design is finished, or result from the fact that the amount of work to be carried out will only be known when a start is made. This sec-ond condition often occurs on maintenance projects.

When work is done it will be paid for at the rates in the schedule and thus, like the bill of quantities, this method requires the work carried out to be measured at intervals. Because the quantities of the various items of work in the job will not be well defined when the Contractor prices them, the prices are likely to be higher than for similar work on a bill of quan-tities contract. The Contractor, not knowing how much of any item he/she will be asked to do, must assume that the quantities will be small and will therefore have to price each item on that basis; that is, at a higher level.

(iii) Cost-reimbursement contracts

Rather than ask the Contractor to give a price for carrying out the work of the project, in this approach the Promoter agrees to pay the Contractor the sums incurred in constructing the job, together with a fee to cover the Contractor's overheads and profit. There will usually be a condition in the agreement that the Employer will not pay for repeating work that was defective as a result of the Contractor's shortcomings, but apart from this, the Contractor takes little risk. When compared with most other contract types, this is a distinguishing feature of cost-reimbursement contracts.

Additional points stemming from this unusual, distinctive method are:

1. Because of the low risk to the Contractor, claims for extra payment should be minimised.

2. The Contractor's accounts must be open for scrutiny to confirm actual amounts to be paid for work done.

3. Any discounts received by the Contractor for bulk purchases should be passed on to the Promoter.

The various categories of cost-reimbursement contract differ principally in the way in which the Contractor's fee is determined. Figure 2.3 illustrates the difference graphically.

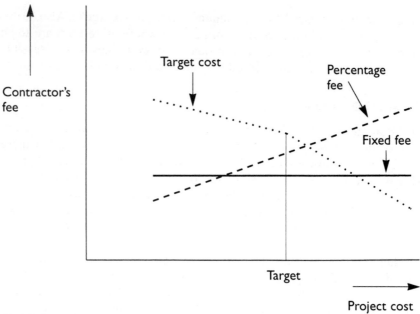

Fig. 2.3 Cost-reimbursement contracts: Contractor's fee

(a) Cost + fixed fee

No matter what the final cost of the project, the Contractor will be paid a fixed, pre-agreed fee to cover overheads and profit. The fee may well have been the subject of negotiation between the Promoter and the Contractor and will be tied to the description of the work to be carried out. If this is changed, there will usually be a mechanism for amending the fee as a result. In general, there is thus an incentive for the Contractor to keep the cost of the job down, for then the fee paid will represent a greater percentage of the work done and the Contractor's overheads are likely to be lower, leaving a greater amount of the fee as profit.

(b) Cost + percentage fee

Here the Contractor's fee varies with the cost of the project. The higher the project's cost, the higher the Contractor's fee. This is seen as an inherent defect as it gives no financial incentive to the Contractor to control costs. In fact, it might be said that maximum inefficiency leads to maximum returns. That said, however, this method may be the only acceptable way of getting the services of a Contractor at short notice; for example, when emergency works must be carried out.

(c) Target cost

An agreement must be reached with the selected Contractor on a 'target cost' for the project, and a fee to be paid if the contract comes in at the target value. A procedure is then agreed for sharing the savings if the project is completed at less than target cost and for allocating the cost overrun if the target cost is exceeded. Thus both parties will gain if the project is completed at a lower cost and both will suffer if the project is more expensive than the target cost. The incentive for the Contractor to work efficiently should be particularly keen.

Different ways of arranging the parties

(iv) Management contract

The traditional method of organising the various parties that contribute to the project is shown schematically in Fig. 2.4, with the Promoter employing an Engineer to design the job and to deal directly with the Contractor. This structure has been the principal way of dealing with projects in the construction industry for a number of years, but has always been recognised as having certain drawbacks. One of these is the fact that the industry is effectively separated into those who design and those who construct, and there is a concern that this may lead to designers who may produce projects that are difficult or particularly expensive to construct. The management contracting approach aims to overcome this problem by making use of the Contractor's expertise during the design process.

Figure 2.5 shows the basic management contracting structure, with the Promoter employing a designer to design the works and a Management Contractor to arrange its construction through Construction Contractors. The Management Contractor should be retained early in the project's life, for he/she is expected to have an input into the design process, helping the designer to produce a design that is easier to construct. When the design is completed, the management contractor then arranges its construction by letting packages of work to other contractors and supervising those contractors on the Promoter's behalf. The Management Contractor does not

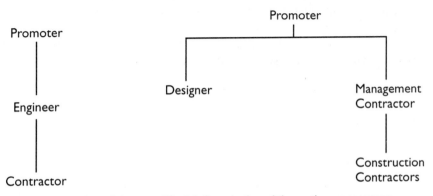

Fig. 2.4 Organisation of the parties: traditional contract

Fig. 2.5 Organisation of the parties: management contract

normally undertake any of the construction work and is paid a fee for the management services provided. The contracts let to the Construction Contractors may be lump sum, measurement or cost-reimbursement contracts.

The main advantages claimed for this method are as follows:

1. The Management Contractor, usually a subdivision of a contracting company, brings construction expertise to the design stage.

2. Construction of the project can commence before the design is complete, as work is let in packages.

3. On complex, multidisciplinary projects where many design groups and many contractors are involved, the contractor's ability to coordinate and control should be invaluable.

4. Time targets should be easier to meet as later packages of work can be adjusted where necessary.

(v) Design and build contract

The Promoter, either directly or through an adviser, states project requirements in general terms and invites contractors to submit proposals and terms of payment for the design, construction and possibly commissioning of the project. This system, shown in Fig. 2.6, has been in use for a number of years in the chemical and oil industries and is becoming more frequently adopted in the civil engineering sector in the UK, with the privatised water companies employing it and the Department of Transport using it in trials of alternative contract arrangements. In this case, the Design and Build Contractor may be a consortium of contractors and consulting engineers getting together in an *ad hoc* agreement to command all the resources required to bid for such projects. The layout shown in Fig. 2.6 represents the contractual system, which is very simple and involves only two parties: the Promoter and the Design and Build Contractor. However, as suggested earlier, the Promoter will usually retain the services of a consultant to advise on design and to monitor and pay for the work as completed. The contract between the Promoter and the Design and Build Contractor will either be a lump sum contract, probably with staged payments, or a cost-reimbursement contract.

Fig. 2.6 Organisation of the parties: design and build contract

(vi) 'BOOT' contract

The acronym BOOT stands for Build, Own, Operate, Transfer and is a quite revolutionary approach to the realisation of construction projects. The change affects both the methods of payment and the organisational structures. The fundamental approach involves what is usually a consortium of financiers, contractors and consultants offering to construct some project, which would otherwise have consumed public money, at their own cost. They recoup that cost over a period of time in which they operate the facility and collect money from the public who make use of the facility. After a pre-agreed period, the ownership of the facility reverts to the general public. Recent examples of this type of arrangement are the Dartford crossing and the Channel Tunnel, and Fig. 2.7 shows diagrammatically, the parties involved. The concession company promotes the project and may acquire funds for its construction via banks and through equity investors. Design and construction are carried out under the construction contract and the operator (for the Channel Tunnel, this is Eurotunnel, which is also the concession company) operates the facility to recover money from the users.

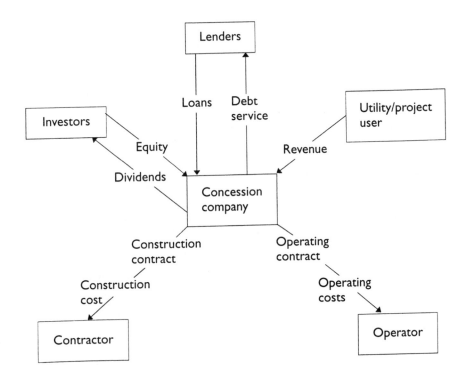

Fig. 2.7 Organisation of the parties: BOOT contract
Source: G. Haley (1972) Private finance for transportation and infrastructure projects. *International Journal of Project Management* 10(2).

Recommendations for further reading

The Construction Industry Research and Information Association (CIRIA) has produced a number of useful documents describing the different procurement methods used in the construction industry. The publications available are:

Report 85:	*Target and cost reimbursable construction contracts,* J.G. Perry, P.A. Thompson and M. Wright
Report 100:	*Management Contracting*
Special Publication 113:	*Planning to build?,* M. Potter

3 Contract documents

Introduction

Contract documents in civil engineering will be drawn up for each particular project and will include, as a minimum, a definition of what is to be constructed and also of how it is to be paid for. In the traditional arrangement of the parties, it is the Engineer who draws up the documents and the Contractor who gives a price for completing the work detailed in them. As shown in Fig. 3.1, preparation of the documents is the culmination of the design process. Having completed the calculations and decided what

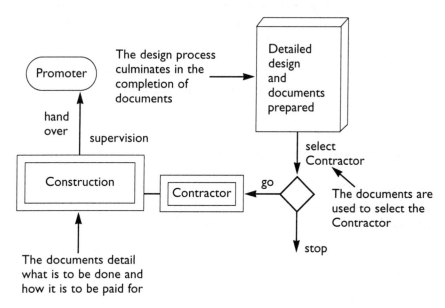

Fig. 3.1 The importance of contract documents

is to be constructed and from what materials and to what quality, the designer must then specify these details in the contract documents.

When selecting a Contractor, the documents also play an important role. All of the contractors who are asked to tender for the project will be sent a copy of the documents, and with this understanding of what the project entails, they will give their price for carrying out the work. It is then clear that it is only what has been laid down in the documents that the contractor can be expected to construct for the price tendered.

These documents are also of fundamental importance in the construction stage of the project, defining as they do, exactly what is to be done and how payments are to be made. When the contract is under way, if any of the details are wrong, or the Promoter wants to change them, this can usually be done, but will typically require the payment to the Contractor to be revised from the price offered at tender. Whenever there is any disagreement about exactly what is to be constructed, to what quality or how it is to be paid for, it is the contract documents that will be consulted first to attempt to decide the issue.

Having sent the documents to the contractors at the beginning of the tender stage, there is then a period in which the contractors pore over them to make sure that they fully understand what they are expected to do before giving a price for carrying out the work. During this process, deficiencies in the documents may come to light and contractors may notify the Engineer of any deficiencies they find. This may lead to the Engineer realising that certain details should be changed and, if so, all contractors tendering will need to be informed of any such changes. They will then all be pricing a revised set of documents that better describe the work to be carried out. This means, however, that the documents have undergone a transformation, and this is usually recognised in the industry by calling the set of documents sent out to tender the *tender documents* and the set that the contractors finally price, the *contract documents*. If no changes were made during the tender stage, there will, of course, be no difference between the two sets.

What contract documents are

As detailed in Chapter 1, with a traditional arrangement of the parties and where a bill of quantities measurement-type contract is being used, the documents normally produced will consist of the following:

1. form of tender

2. form of agreement

3. conditions of contract

4. specification

5. bill of quantities

6. drawings.

It is the drawings and the bill of quantities that will normally take the longest time to complete, as these must be produced uniquely for each project. The form of tender and the form of agreement are quite straight-forward, and standard versions are available that can be used with little effort. Also, standard conditions of contract and specifications exist and these can be incorporated, with or without amendments, simply by naming them; for example, it might be stated that 'The Conditions of Contract referred to in the Tender shall be the *Conditions of Contract* (sixth edition) (1991), commonly known as the ICE Conditions of Contract.' If the Engineer recommends that a standard document be used, but that some amendments should be made to it, this statement would be followed by a list of sections of the standard document that are to be modified and would list the modifications.

Another document that is usually produced and included with the other documents sent to the tendering contractors is the 'Instructions to Tenderers'. It is not a contract document and thus would never be consulted to help to define the work involved in the contract. It is simply a guide to the contract documents and helps the tendering contractors understand what they must do to submit an acceptable tender. For example, where a bill of quantities has been used, a statement may be made in the 'Instructions' that the bill must be fully priced, meaning that a tender that does not put rates against the various items in the bill, but simply gives a lump sum price, will not be accepted.

Each of these documents will now be described in turn. The coverage of Bills of Quantities and Conditions of Contract in this chapter will be limited, as whole chapters later in the book are devoted to these two documents.

Form of tender: the Contractor's written offer to carry out the works

The form of tender is usually a letter, from the Contractor to the prospective employer, agreeing to carry out the works for a sum of money. On a lump sum contract, this sum might actually be stated, but on a bill of quantities contract, the wording is more likely to be, 'for a sum which is to be determined in accordance with the Conditions of Contract'. The calculation of exactly how much the Contractor should be paid on a measurement contract is not a simple matter, as will be seen later. There is usually an appendix attached to the form of tender, and here the Engineer will provide some general information about the project, such as:

- the amount of insurance the contractor must take out;

- the time allowed for completion of the works;
- amount of liquidated damages;
- minimum amount of interim certificates.

More explanation of these items is given in Chapter 7.

Form of agreement: the formal agreement between the Promoter and the selected Contractor

The form of agreement lays down the agreement between the two parties to the construction contract. It defines the documents that are to be read as part of the agreement and should be signed by both parties. However, a valid contract is formed by the Contractor's offer (form of tender) and a letter from the Promoter accepting the offer. The form is therefore not always completed.

Conditions of contract: the rules by which the contract will be run

The conditions of contract document relates to the contract between the Promoter and the Contractor to construct the works. The document is likely to be very complicated, and as has already been said, standard versions are available which can be adopted. In essence, it identifies the parties involved in the contract and defines their roles and responsibilities. It will also lay down how the Contractor is to be paid and will allocate responsibility for some of the construction risks that are known to occur frequently. The document effectively governs how site personnel should conduct themselves, and any engineer intending to work on site should therefore have a working knowledge of the conditions of contract being used on that site. For civil engineering work undertaken in the UK, it is the ICE *Conditions of Contract* (6th edition) that are likely to apply, and for this reason it is this particular form of contract that will be dealt with in the subsequent section. All Conditions of Contract will, however, cover similar material, and if the engineer has been exposed to one set of conditions, understanding another set will be made easier as a result.

Specification: a detailed statement of what is required in terms of the quality of the materials, the methods to be used and the workmanship required

It is common to use a standard text, such as the *Specification for Highway Works*, which covers a variety of types of work, including fencing and guardrails, earthworks and bridge expansion joints. The standard form may be modified where the Engineer has misgivings about parts of the standard and may be added to where the project includes work that is not covered by the standard. It is, of course, possible for the Engineer to write his/her own specification, but the in-depth knowledge needed for such an enterprise will usually make it prohibitive in terms of time and money to do so. Some organisations do, however, develop their own clauses for particular areas of work which they will want to adopt for any project they are involved in.

Examples of the different types of information given in the *Specification for Highway Works* will now be shown, to give the reader some appreciation of what can be expected to be found there.

Quality of materials: Clause 804 defines sub-base material type 2, as natural sands, gravels, crushed rock, crushed concrete or well burnt non-plastic shale. It also requires the material to comply with:

- a grading envelope;
- a CBR value;
- a range of moisture content when laid;
- a 10% fines test to BS 812;
- a soundness test to BS 812.

Methods to be used: Clause 608, which deals with the construction of fills, provides considerable instruction as to how this is to be done. Some of the details are:

- embankments shall be constructed evenly over their full width and constructional plant and traffic shall be directed uniformly over them (where a haul road through the site traverses the fill, this means that the position of the haul road must be changed at intervals);
- fill material is to be deposited in layers, whose depth depends on the material itself and the compaction plant being used;
- when fill is to be deposited against an existing slope, the face of that slope must be benched before the fill is deposited.

Workmanship: Clause 702 defines the tolerances to which road pavements must be laid. Horizontally, the pavement must be constructed within ±25 mm of the design position. Vertically, a table of tolerances is given, in which the values depend on the layer of road pavement being considered. Sub-bases must be laid within +10 mm and -30 mm of the design level, while road surfaces must be laid within ±6 mm.

Bill of quantities: a quantified list of items of work which together represent the project to be constructed

The bill of quantities is compiled in accordance with a set of rules called a *method of measurement*. These rules govern how the work of the project is to be broken down into items whose quantities will then be measured and placed against each item. The bill of quantities is then a breakdown of the work of the job that the contractor can rely on as properly representing the project. To produce an acceptable tender, one of the things the contractor absolutely must do is to put a price against each of these items. When the price of each item is multiplied by its quantity, and all of the sums thus calculated are added together, the result is the major part of the contractor's tender.

Drawings: the essential statement of what is to be constructed

Drawings provide the most accessible method of communicating information in civil engineering, and in most situations will be the first document consulted. Each organisation will have its own standard method of producing drawings, with a unique numbering system and standard title blocks and revision blocks. As work is completed, and when changes have been made, the relevant drawings should be amended to reflect the changes and the drawing number updated to indicate that the drawing has been revised.

Certain details will be required on more than one drawing (e.g. types of fence, gates, kerbs, etc.), and these may be fully specified on separate 'Standard Detail Drawings', which may be called up on the main drawings as required. This permits much visual information to be standardised, but there will still need to be main scheme drawings produced to show the layout of the various parts of the project.

It will often be necessary to produce more than one drawing of a particular part of the contract, in order to sensibly show all the different aspects of the work. Thus for a roadworks contract, as shown in Fig. 3.2, a number of drawings will be needed to show the whole of the project to a suitable scale and for each of those drawing areas separate drawings may be needed to show:

- site clearance

- setting out

- general layout (incl. drainage, fencing, kerbing, etc.)
- traffic signs and road markings
- public utilities' plant and diversions.

There will also need to be additional drawings showing:

- cross-sections
- longitudinal sections
- typical cross-sections (showing details of construction, dimensions and layout at a typical cross-section).

To understand what is to be constructed, the Contractor will often have to look at a number of documents together. Figure 3.3 gives an example of the need to consult the scheme drawings, standard detail drawings and the specification to fully appreciate the where, how and what of laying kerbs. The scheme drawings show where a particular type of kerb is to be laid, the standard detail drawings show the dimensions of the kerb and the bed and backing details, while the specification details material qualities and laying tolerances. As the Conditions of Contract say, the contract documents are 'mutually explanatory'.

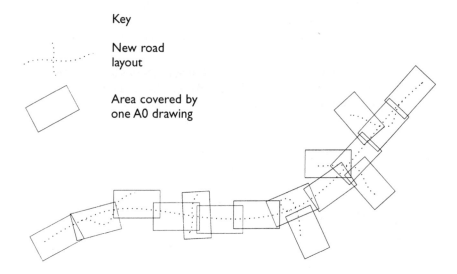

Fig. 3.2 Plan drawings for a roadworks contract

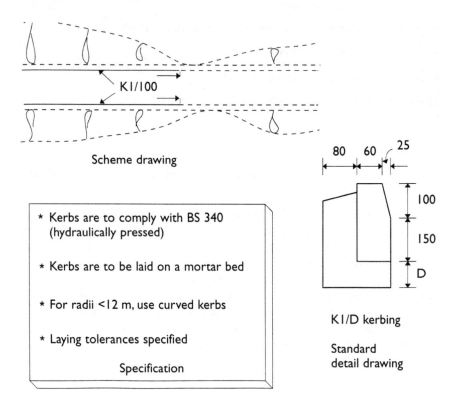

Scheme drawing

* Kerbs are to comply with BS 340 (hydraulically pressed)

* Kerbs are to be laid on a mortar bed

* For radii <12 m, use curved kerbs

* Laying tolerances specified

Specification

K1/D kerbing

Standard
detail drawing

Fig. 3.3 Kerbing: documents to be consulted

Recommendations for further reading

The specification is an important document and the reader may wish to inspect a copy of a standard version, for example:

Department of Transport (1991) *Manual of Contract Documents for Highway Works*, Volume 1: *Specification for Highway Works*, HMSO, London.
The bill of quantities and the conditions of contract are described in more detail in Chapters 4 and 7.

4 Civil Engineering Standard Method of Measurement, 3rd edition: CESMM3

Introduction

Bill of quantities measurement contracts are very common in many areas of civil engineering, and for this reason it is important to have a good understanding of how a bill of quantities is compiled and how it is used throughout the project. The first thing to recognise is that a bill of quantities will be prepared in accordance with a set of rules, known as a method of measurement. Depending on the type of work involved in the project, different methods of measurement are available, including the following:

Civil Engineering Standard Method of Measurement, 3rd edition	for civil engineering projects
Method of Measurement for Highway Works	for road and bridgeworks projects
Standard Method of Measurement for Building Works, 7th edition	for buildings

It is *Civil Engineering Standard Method of Measurement,* 3rd edition, or CESMM3, that has the most general application in civil engineering, and this is the method of measurement that will be described here. However, before we look at the detail of CESMM3, there is more that needs to be understood about bills of quantities themselves: how they change as the project progresses and the reasons why it has been found necessary to develop sets of rules for their production.

Stages of development of the bill of quantities

The main uses of bills of quantities are explained in Chapter 2, from which it should be clear that a bill of quantities is an evolutionary document passing through a number of stages. The main stages are as follows:

1. When first produced, the bill will contain estimates of the quantities of each of the bill items but no rates against these items. In this condition, the bill is sent out to the contractors tendering for the project.

2. Having been priced by the contractors, there are now rates against each item, and mainly on the basis of the price information the Engineer will assess the various tenders and recommend which contractor should be selected.

3. The selected Contractor will begin work on the job and each month will be paid for the work done to date. To achieve this, the Engineer's site team will tour the site each month to find out exactly how much work the Contractor has completed, and these amounts will be recorded against the relevant bill items. Also, if any additional work has been added to the project that is covered by a bill item, this may be paid for alongside the original contract work. Thus, for each month of the contract, there will be a version of the bill of quantities with increasing amounts against the items as the project progresses.

4. When the work is complete, the final quantities for each bill item should have been agreed between the Contractor's and Engineer's site staff. This revisiting of each bill item to remeasure the amounts of work carried out confirms the fact that the quantities in the original bill of quantities are just estimates. The reassessment of quantities is known as 'admeasurement'.

Need for a method of measurement

The bill of quantities is not intended to describe the nature or extent of the contract work, but rather to be a system for offering prices to the Promoter for carrying out the work. To do this effectively, a standard approach is needed and this may be best appreciated by considering the consequences of compiling bills without such a standard. In that event, engineers would have no guidance about the level to which the work of a contract should be broken down. Some might consider that a reinforced concrete element should be dealt with as a single item, whereas others would want to provide separate items for formwork, reinforcement, providing the concrete and placing it. For work on laying road pavement, some might measure this using units of area while others might want to measure tonnages of material laid. Thus, without standardisation here, there are opportunities for a variety of philosophies to flourish. To make the case for a method of measurement it is therefore necessary to make the case for standardisation – the main consequence of using a method of measurement. The most important advantages gained from standardisation – which means that two separate engineers will produce virtually the same bill of quantities for the same project and that stock item descriptions will be used to describe the same type of work, are:

1. Contractors will become familiar with the work breakdown used and will quickly understand the work represented by each item. The need for speed is vital when contractors must price a complicated project in a limited time (the tender period).

2. Rates from previous tenders will be held by the contractor and for the same bill item may be used as a guide to the rate to be offered for the work in a tender currently being priced.

The main sections of a CESMM3 bill of quantities

The discussion so far has concentrated on what is known as the 'work items' section of the bill of quantities, but CESMM3 makes it very clear that a number of other sections must be produced as part of a complete bill of quantities. These are:

- list of principal quantities
- preamble
- daywork schedule
- work items (grouped into parts)
- grand summary.

Each section will now be dealt with, in turn.

List of principal quantities

The list of principal quantities is a list of the main components of the job together with the approximate quantities. It is given to help the tenderers gain a quick understanding of the scale and character of the project. Items in this list are likely to be amalgamations of the main areas where items are provided in the 'work items' section. For example, there may be a number of items in the main bill describing reinforced concrete work and these would be represented in this list as a single general item for such work with one combined quantity.

Preamble

In the preamble, the method of measurement used to prepare the bill is declared together with any amendments to the standard document. It is quite acceptable to use CESMM3 for a contract but to wish to amend parts of it or to add other parts necessary to provide a better description of the contract work. Any amendments or additions are included in the preamble. For some projects, different methods of measurement may be used for different parts of the work, and where this is the case it must be acknowledged.

A definition of rock may also need to be given here if the contract includes any excavation, boring or driving. It is invariably more difficult to work in rock than other materials, and to reflect this difficulty, separate items will be provided in the main bill (say) for the excavation of rock and the excavation of other less exacting materials. This allows the contractor to give a higher price for the more difficult work, but it is essential that rock be defined in some way so that the engineers on site can distinguish the quantities to be paid at the different (often considerably different) rates.

Daywork schedule

There will invariably be instances during the construction of the work when changes to the original details will have to be made. One way of paying for these changes is to record the actual amounts of labour, plant and materials used in making the changes and to cost up these amounts at agreed rates. The daywork schedule is the list of rates that are to be used in such circumstances. CESMM3 recognises two ways of defining these rates. Either the Engineer produces a list of the various classes of labour, plant and materials likely to be needed and asks the Contractor to provide rates for each, or use is made of a standard, nationally produced and regularly updated set of rates. This will be the 'Schedules of Dayworks Carried Out Incidental to Contract Work', issued by the Federation of Civil Engineering Contractors. This is the easiest and often the best method, and the rates can be tailored to the particular contract by requesting the contractor to quote adjustments, either additions or deductions, to the rates.

Work items

The 'work items' section is by far the most complicated and detailed part of the bill of quantities, and although an attempt will be made here to introduce the essential elements, the best way of understanding work items is to actually produce a bill for a project.

As the title implies, it is in this section that the main breakdown of the work of the project into standard items occurs. There are 26 separate categories, known as 'work classifications', under which the work must be measured, ranging from Class A: General Items, through to Class Z: Simple Building Works. One of the first tasks facing the engineer compiling the bill is to decide in which classes items will be needed. For example, if there is in-situ reinforced concrete work in the job, it is likely that Class F: In-Situ Concrete and Class G: Concrete Ancillaries will definitely be needed. Class F deals with the provision and placing of concrete, while Class G deals with formwork, reinforcement, joints and finishing of surfaces. Once the classes where items will be needed have been recognised, the next step is to generate the actual item descriptions themselves.

However, before proceeding to the formulation of item descriptions the Engineer must recognise the possibility that this work items section of the bill may be split up into separate parts to expose cost differentials arising from special circumstances. This is perhaps best explained by reference to Fig. 4.1, which shows the hierarchical organisation of a set of documents for a major road and bridgeworks contract. The bill of quantities is clearly one of the contract documents, which in turn is composed of a number of

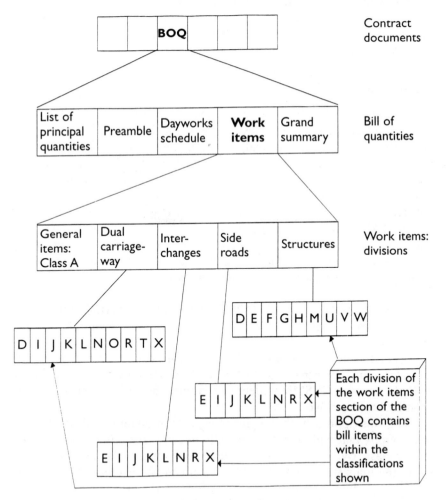

Fig. 4.1 Hierarchical structure of a bill of quantities (BOQ)

sections including the work items. These are shown as subdivided into class A items and then separate parts to cover the main dual carriageway, the interchanges, the side roads and the structures. The project has thus been broken into separate geographical areas and each area will be treated like a separate bill of quantities. In each of these separate sub-bills there will typically be a number of items that are common; for instance, items for kerbing, earthworks, etc. This increases the complexity of the bill, but is seen to be worthwhile because, as stated above, it exposes cost differentials. It is considered likely that the cost of carrying out work on the dual carriageway will be different from the cost of the same work on (say) the side roads. This is because on the dual carriageway, there will be long lengths to work at and productivity should be high with a consequent reduction in cost. On the side roads, where much shorter areas are to be worked on and there is the complication of tying in to existing kerbs and dealing with traffic, productivity is likely to be lower. Thus by providing separate items for similar work in different areas, the contractor will be better able to provide rates that relate more closely to the cost of carrying out the work described.

Formulating item descriptions

Each of the 26 classifications in CESMM3 contains details of what is included and excluded in the class; a systematic listing of elements of item descriptions; and notes to provide additional explanation. Having decided that items will be needed in a particular class, by checking the included/excluded section of that class, the engineer compiling the bill should then read the explanatory notes for the class before formulating the descriptions. To produce an acceptable item description conforming with CESMM3, the engineer must study the element listings and will normally create items by selecting one element from division 1 and combining it with an element from division 2 and one from division 3. For example, in Class G, a formwork item might be:

FIRST DIVISION	SECOND DIVISION	THIRD DIVISION
Formwork rough finish/	plane vertical	/width 0.1–0.2 m

It is clear that formwork with different finishes cannot be combined into the same item, nor can vertical formwork be merged with battered formwork. In this way, CESMM3 effectively defines the level at which the work must be broken down.

Using this approach, the engineer must then generate all of the item descriptions that will be necessary fully to cover the work of the project in this class. By repeating the exercise in all of the classes where work exists, the main work items for the project can be developed. The quantities for each item must then be 'taken off' from the drawings and included against the relevant item description. The following notes, while not providing all the additional information needed to carry out the identification and measurement of items, should give an indication of some of the complexity:

1. It is not necessary to say, 'supply, deliver, cut, bend and fix mild steel bar reinforcement to BS 4449'; this is understood by the description, 'mild steel bar reinforcement to BS 4449'. This approach applies generally in CESMM3, where components, not tasks, are identified.

2. Units of measurement for each particular work item are specified *and are to be used.*

3. Where the description of an item is not fully covered by the three divisions in CESMM3, it is acceptable to add additional description. Equally, one or more of the descriptions in the divisions may not be needed and may be replaced by reference to the drawings or specification.

4. Any item of work that does not fit readily into CESMM3 may be fully described in the contract and a description included in the bill against which the contractor is required to state a sum.

5. Quantities are to be computed net from the drawings with no allowance for shrinkage or wastage. It is normally acceptable to round up quantities to the nearest unit, with the exception of such items as site clearance (unit: hectare) and reinforcing steel (unit: tonne), where decimals should be used.

6. Because of the likely effect on pricing, all items of work affected by water should state this in the item description.

7. For work which involves excavation, boring or driving, a number of surfaces are defined. These are the Original surface, the Commencing surface, the Excavated surface and the Final surface as shown in Fig. 4.2. Where the Commencing surface for an item is not the Original surface (as for items B and C), then the Commencing surface must be identified in the item description. Where the Excavated surface is not the Final surface (as for item A), then the Excavated surface must be identified in the item description.

Coding and numbering of items

CESMM3 does not insist that items be coded, but provides a useful system for doing so, which it is recommended should be adopted. The system stems from the listing of elements of item descriptions. For example, the code G112 identifies the item under concrete ancillaries: Formwork rough finish plane vertical width 0.1–0.2 m, and is derived as follows:

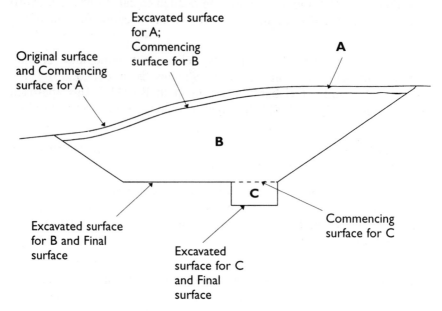

Fig. 4.2 Earthworks surfaces

Class	G	concrete ancillaries
Element selected from the first division	1	formwork: rough finish
Element selected from the second division	4	plane vertical
Element selected from the third division	2	width 0.1–0.2 m

It is made very clear that these item codes do not have any contractual significance; it is the written item description that should be consulted to understand what is to be priced. This means that a bill of quantities containing only codes without descriptions would not be in accordance with CESMM3. A number of variations are permitted, as follows:

1. Item code B204 signifies that nothing has been selected from the second division.

2. Item code G24* relates to all items ranging from G241 to G245.

3. F249 means that a feature has been used in the third division position that is not listed in CESMM3 (which has a maximum of eight selections in any division).

4. R323.1 means that additional information beyond what is provided by the three divisions has been used in the item description.

Class A items

Items under the Class A heading are quite different from the normal work items to be found in classes B to Z. They are known as General items and comprise contractual requirements, preliminaries, method-related charges, provisional sums and prime cost items. As shown in Fig. 4.1, there would only need to be one part of the bill including Class A items, irrespective of how many parts the bill had been broken into.

Preliminaries

The preliminaries include items such as site accommodation for the Engineer's staff, transport and surveying equipment that the Engineer wants the Contractor to provide. Items covering accommodation and services for the Contractor would not be included by the Engineer, but might be added by the Contractor as 'method-related charges'.

Method-related charges

Under the heading of method-related charges the Contractor may insert his/her own items to cover certain costs which are not covered by the bill items included by the Engineer. These might include the Contractor's own site accommodation, as previously stated, the cost of temporary works and supervision costs. Such items must be included either as fixed charges, e.g. the provision of a temporary access road, or as time-related, e.g. supervision costs.

A traditional bill of quantities tended to measure only the permanent works which were left behind when the Contractor's men and machines left site. In practice, a good deal of the Contractor's costs are tied up in temporary works and resources which are not easily reflected by the quantities of permanent works constructed. When the works have to be changed, there will often be an effect on these 'indirect costs' which is not taken into account by simply pricing the extra work at a rate similar to one which already exists in the bill. If the Contractor has included such indirect costs as method-related charges, then these can be taken into account when the varied work is priced. As an example, consider a site that is split in two by a river. Excavation carried out on one side of the river will be used as fill on the other side, and for this reason the Contractor will install a temporary structure to allow earth-moving vehicles to cross. If the temporary bridge is hired, then the longer it is in use, the more the Contractor must pay. Should the Contractor be instructed to carry out additional earthworks by the Engineer, which requires the bridge to be in place longer than initially planned, the Contractor can reasonably expect to be paid for excavation at a rate similar to the one in the bill of quantities, but also for the extended use of the bridge. If the Contractor has included the temporary bridge rental as a method-related charge, there should be little room for argument.

Prime cost items

Prime cost items usually relate to work carried out by a 'nominated sub-contractor' (NSC) and should have a sum quoted against them by the Engineer. A nominated subcontractor is a contractor selected by the Engineer, usually to carry out specialist work on the site occupied by the main Contractor. The main Contractor is expected to co-operate with the NSC as with a normal subcontractor. The sum quoted will be the prime cost of the work to be carried out by the NSC as agreed between the Engineer and the NSC. The main Contractor can include a sum for providing facilities for the NSC and can also include a sum for profit.

Provisional items

Provisional items are sums of money included in the bill of quantities to cover contingencies (the unexpected), and may be included in Class A to cover a specific outcome or in the grand summary (*see below*). The Contractor will not be paid these sums, but they will be available if needed, and if the Contractor carries out additional work he/she will be paid as much as is deserved for the extra work done. Effectively, the provisional sums represent an expectation that the cost of the work may increase beyond what is covered by the normal bill items and make additional money available.

Recommended procedure for preparing the work items section of a bill of quantities

1. Decide on any special division of the bill into parts as a result of location, access, timing, etc., to expose cost differentials. Determine exact demarcations between these sections.

2. For each part in turn decide which work classifications are needed to cover the work to be measured.

3. Within each work classification read the notes and decide what item descriptions will be used for measuring the work.

4. Take off quantities for each item in turn using dimension paper (*see* Fig. 4.3), and transfer the quantities to the bill paper.

5. Consider Class A items including prime cost items and provisional sums, and and allow an opportunity for the contractor to insert method-related charges.

6. Draw the elements of the bill together. This will involve providing class summary sheets, part summary sheets (where the bill has been divided into parts) and a grand summary sheet.

Taking off quantities

The following notes refer to Fig. 4.3:

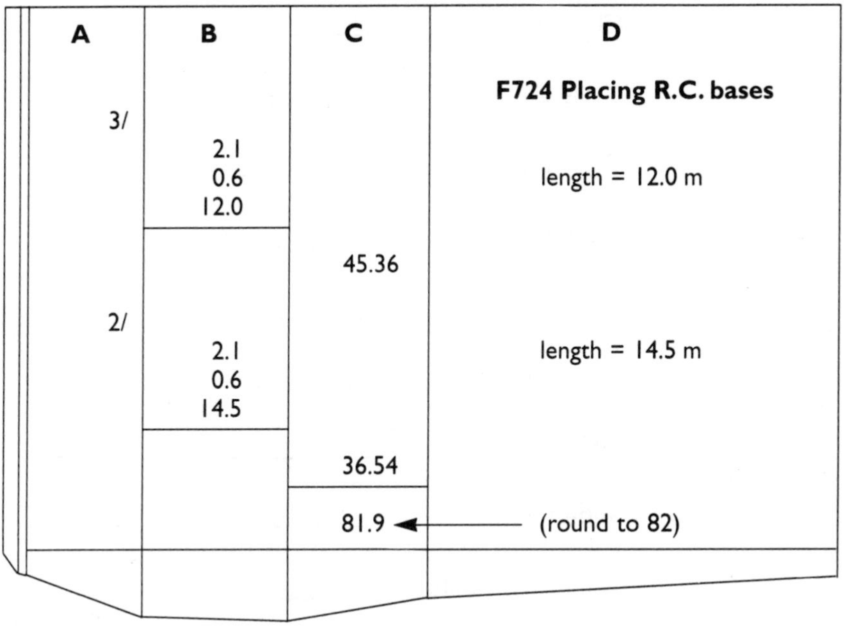

A	B	C	D
			F724 Placing R.C. bases
3/			
	2.1		
	0.6		length = 12.0 m
	12.0		
		45.36	
2/			
	2.1		length = 14.5 m
	0.6		
	14.5		
		36.54	
		81.9 ◄———	(round to 82)

Fig. 4.3 Take-off sheet

Column A   The 'timesing' column; where there is more than one element with the same dimensions, this column is used to multiply the dimensions by the number of elements.

Column B   The 'dimension' column where the measurements taken from the drawings are set down.

Column C   The 'squaring' column contains the volumes, areas, etc. derived from dimensions in columns A and B.

Column D   The 'description' column describes the work to which the dimensions apply.

Grand summary

This final section of the bill of quantities is the part in which the totals from each of the divisions of the bill are brought together to a grand total for the whole contract. At its simplest it will look like this:

	£	p
Class A	_____	___
Class B	_____	___
Class C	_____	___
Class E	_____	___
Class F	_____	___
Class etc.	_____	
Sub-total	_____	___
GCA	_____	
Subtotal	_____	___
Adjustment item	_____	
GRAND TOTAL	=========	

For each of the classes used, a number of pages may have been needed to accommodate all the items necessary to measure the contract work. On each of these pages, space should be made available for the contractor to insert a page total when all the rates have been included and the extensions calculated. A class summary sheet should also be provided where all the page totals in a particular class can be brought together. The class totals are then brought to the grand summary as above. Figure 4.4 illustrates this process. Where the bill has been subdivided into parts, to expose cost differentials, class summaries will be brought to part summaries and then the part summaries will be brought to the grand summary.

The general contingency allowance (GCA) is a provisional sum which may be inserted by the Engineer and which operates in the same way as the provisional sums in Class A. The adjustment item is an opportunity for the Contractor to make an addition to or deduction from the grand total without amending each of the rates in the main bill.

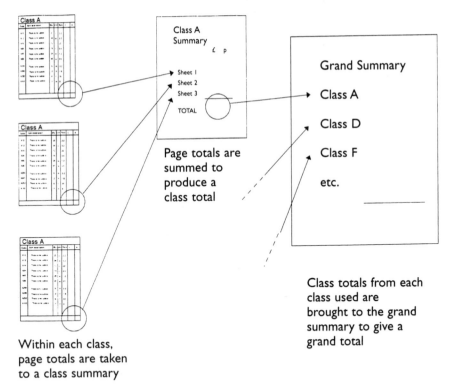

Fig. 4.4 Use of summary sheets

Recommendations for further reading

For a good understanding of the way in which a bill of quantities is produced, the reader should ideally use CESMM3 to produce a bill for a small contract. At the least, the document itself should be studied and it would also be valuable to consult the *CESMM3 Handbook*.

Institution of Civil Engineers (1991) *Civil Engineering Standard Method of Measurement*, 3rd edition. Thomas Telford, London. 109 pp.

Barnes, M. (1992) *CESMM3 Handbook*, 2nd edition. Thomas Telford, London. 235 pp.

5 Planning and control

Introduction

All plans involve attempts to predict the future. Sometimes the plan produced will depend on predicting how some variable over which we have little or no influence is likely to change and then making decisions about our own activities. For example, if our company manufactures a particular product, the predicted demand for that product over the next five years will be a very important factor in deciding our future actions. If demand is thought likely to rise, we might consider increasing our production capacity by, say, building new factories and warehouses. If demand is likely to be stable, we might retain our existing establishment, and if demand is likely to fall, we might consider diversifying into other areas. Most of these courses of action will involve increased costs, and the money, which may have to be borrowed, is at risk if our predictions turn out to be incorrect.

Other types of plan involve predicting how well we will be able to perform in specific circumstances in the future. If we are to offer a price for doing work, we must have a good understanding of what our costs will be to be sure that any percentage added to those costs for our profit is actually realised in practice. This kind of plan will require an assessment of the resources to be used on each part of the work, the likely productivity of those resources and from this information a prediction of our likely costs. In such a plan there are also risks. If the costs of doing the work are underestimated, the anticipated profit may not be realised; we may even make a loss.

In this book, it is the second type of plan that will concern us most, but before getting into the detail of what I will call *project planning*, a little more time will be spent considering the first type of planning. Where this is an attempt to predict the company's prospects in the future and to make decisions on that basis, this is often known as *policy planning*.

Policy planning

All companies and organisations that must earn income to continue to trade will typically have exactly that as one of their prime objectives: to continue to earn income and stay in business. However, the company will also usually have other objectives. These might involve increasing its share of the markets in which it operates or diversifying into other markets. The decisions that must be taken to pursue any of these objectives must be well researched if the risk of failure or financial loss is to be minimised. As suggested in the introduction to this chapter, predictions of the demand for our product(s) will be essential information on which to base our future strategies. In some situations, however, a recognition of the way in which the markets in which we operate are changing may be fundamental to our continuing in business. If the service we offer or the products we make are likely to be superseded by other, new ways of providing that service or by other products, we need to be aware of that impending change as soon as possible. This will then allow us to adjust our operations accordingly. An example can be drawn from the current construction market, where there is a move away from the traditional method of procuring projects towards *design and build* and *BOOT-type contracts*. This is a substantial change. Major civil engineering consultants and contractors must surely be wondering what actions they should take to ensure that they obtain a share of the work available sufficient to allow them to continue to operate, hopefully at the same level and preferably at a higher level of market share.

This kind of planning must clearly be carried out for the company as a whole, with regular reassessments being made of existing operations and markets to ensure that any actions taken are attuned to current and predicted future circumstances and that necessary actions are taken when circumstances dictate.

Project planning

Most engineers in the construction industry are involved in the design and construction of new works or in the maintenance of existing facilities. They will be employed by a client organisation (such as a water authority), a design organisation or a contractor, and the majority will be associated either with one project or with a number of projects. The use of the word *project* here means a complex package of work for which a particular organisation is responsible. The design organisation's project is the design of the works, possibly also including supervision, whereas the contracting organisation's project is its construction, and both will typically have agreed to be involved in the work for an agreed payment. Having agreed the basis on which payment is to be made and typically having a definite time within which the work must be completed (otherwise damages may have to be paid), the consultant or the contractor must work efficiently if they are to make money from the job. Whenever we want to make sure that our performance is good. in many areas of work, we practise and learn from the experience. Such an approach is effectively possible in the process industries, where mass production means that we can sample some of the output, discard it if it is unacceptable and correct any failings so that subsequent products are acceptable. This approach will not work with projects, as there is only one product, which must therefore be properly executed the first and only time it is done.

Why do we plan?

This one-off nature of projects is probably the main reason that special project planning techniques have been developed. Although we cannot practise our project, we can develop a conceptual model of how we expect the work to progress. The various methods of project planning are different ways of modelling the activities which together make up the project. It has been said that the planning process for projects is like *building the project in our minds*, and by doing this we get as close as possible to practising the project's execution. However, it has also been said that *the only thing we can be sure of about our project is that it will not progress in accordance with the plan*. This might be seen as a convincing argument for not bothering with a plan and simply doing our best as the work progresses. However, writers on this topic have tried to convince us of the value of project planning, as the following extracts will confirm:

Present decisions affect both present and future actions and if immediate short-term decisions are not taken within the framework of long-term plans, then the short-term decisions may effectively impose some long-term actions which are undesirable but inescapable. (unknown)

Planning is an administrative process ... necessary so that instructions may be issued in order to instigate action for the achievement of a specific objective(s). (Roy Pilcher)

Control is impossible without a plan. (Michael J. Jackson)

These statements, although very wise, may not be particularly easy to understand and a relevant example will hopefully shed some light. If we look at a common situation in which students might develop their own personal plan, a revision timetable, some of these ideas should become clearer.

Consider a student with four examinations to take at the end of the first semester. Figure 5.1 shows the time-scale, indicating vacation time, a reading week and two weeks in which examinations will take place. The other weeks in term time will contain lectures and tutorials. The following points need to be recognised:

1. The student's main objective is assumed to be to turn up to the examinations as well prepared as possible and to perform to the best of his/her abilities.

Week no.	Starting	SUN	MON	TUES	WED	THUR	FRI	SAT
6	6/11							
7	13/11							
8	20/11							
9	27/11							
10	4/12							
11	11/12	VACN	VACN	VACN	VACN	VACN	VACN	VACN
12	18/12	VACN	VACN	VACN	VACN	VACN	VACN	VACN
13	25/12	VACN	VACN	VACN	VACN	VACN	VACN	VACN
14	1/1	VACN	VACN	VACN	VACN	VACN	VACN	
15	8/1							
16	15/1							
17	22/1		reading	reading	reading	reading	reading	
18	29/1		EXAM	EXAM	EXAM	EXAM	EXAM	
19	5/2		EXAM	EXAM	EXAM	EXAM	EXAM	

Fig. 5.1 Revision plan: vacations, reading week and exam period

2. Other work must be done during this period, including coursework and attendance at lectures and tutorials.

3. Information about the timing of the examinations and the coursework to be done may not all be available at the time the plan is first developed.

4. There is often an unwillingness to produce such a plan as it makes clear just how much work there is to be done!

Decisions must then be made about the following:

- when to start revision;

- what subjects to study when;

- methods of revision;

- how much time is needed for each subject and how much revision can sensibly be done in a day;

- for subjects covered early, how to bring that material back to mind before the exam;

- the balance of effort between revision and coursework.

The resource being planned here is clearly the student's time, and different students will have quite different views as to how much time is needed. Any solution offered is thus for demonstration purposes only and certainly not intended to be used as a blueprint for success. Figure 5.2 is one possible solution.

Some of the decisions incorporated into the plan are undeniable. For instance, the time between EXAM 3 and EXAM 4 in week 19 will certainly be given up to revising the material in EXAM 4. Other obvious uses of time can also be recognised, but outside of these, decisions on the use of time may be more arbitrary. It may not matter whether the material for EXAM 3 or EXAM 4 should be studied in week 14; the important point here is that a decision must be made. Other aspects of the plan may not be so obvious. Note that for weeks 12–15 inclusive, at the start of the period, revision takes place only on weekdays, while from weeks 16 to 19, all available time will be given up to revision. An experienced student may also recognise that the effort expended in the early part of the revision period will probably be less than that in the time closer to the exam. This fact, if true, will certainly mean that the number of days needed for each subject may be partly dependent on when the subject is to be studied. One other matter needs to be addressed, and that is the method to be used to ensure that the time spent on the material (say) for subject 4 in week 12 is not wasted. Some students find that making concise notes on the subject matter they are revising allows that material to be readily brought back to mind at a later time. It does not matter how this is done, provided the method

Week no.	Starting	SUN	MON	TUES	WED	THUR	FRI	SAT
6	6/11							
7	13/11							
8	20/11							
9	27/11							
10	4/12							
11	11/12							
12	18/12		revise 4	revise 4	revise 4	revise 4	revise 4	
13	25/12		revise 4	revise 4	revise 4	revise 4	revise 4	
14	1/1		revise 3	revise 3	revise 3	revise 3	revise 3	
15	8/1		revise 3	revise 3	revise 3	revise 3	revise 2	
16	15/1	revise 2	revise 2	revise 2	revise 2	revise 2	revise 2	revise 2
17	22/1	revise 2	revise 2	revise 1	revise 1	revise 1	revise 1	revise 1
18	29/1	revise 1	revise 1	EXAM 1	revise 2	EXAM 2	revise 3	revise 3
19	5/2	revise 3	EXAM 3	revise 4	revise 4	revise 4	EXAM 4	

Fig. 5.2 Revision plan: decisions on the use of time

works; the important point is to recognise the problem and to make decisions about how to deal with it.

It will, hopefully, be clear from this example that by looking at the whole project – and here the project is the preparation for success in the examinations – short-term decisions, such as when to start and what to study when, can be made which will not impose undesirable long-term actions. It should also be evident that the process of developing a plan is an administrative one. Here the plan is produced by the planner applying his/her mind and using only paper and pencil. For other types of plan, computer software is available, but the procedure is still administrative in nature.

Under the points to be recognised, number 3, it was said that not all of the information needed to produce the plan might be at hand in the first instance. This simply means that the plan must develop as information becomes available. It will also be seen in the following section, 'Control', that plans will often need updating as the work of the project is executed – yet another reason for considering the plan as a living document.

Control

In many areas of our lives, we must exercise control on a daily basis. Driving a car is a very sophisticated example of control. The driver is continually taking in information from traffic signs, about the position of his/her car, other cars and pedestrians, and adjusting speed and position on the road on the basis of that information. Decisions about how much we can afford to spend on eating, drinking and holidays will also, hopefully, be subject to some control. With a knowledge of our income and outgoings, we should be able to make sensible judgements about these matters.

In business, the need for control is also indisputable. The managers who are responsible for carrying out work must be able to answer the question, 'how is it going?', which usually means, 'will we complete the work in time?', 'will we remain within budget?', and, all-importantly, 'will we make money on it?' To be able to give an answer, the managers must have some system of reporting on the progress of the work and some yardstick against which that progress can be measured. However, the only way to try to ensure that the answers to these questions are all affirmative is to be in control of the work and make the necessary decisions as the work progresses. Having monitored progress and found that all is not well, the managers must consider what actions need to be taken to rectify the situation and implement those actions. Only when such a procedure is enacted at regular intervals can we say that we are controlling the work. Many writers believe that control is management's most important function, and the statement made previously that 'control is impossible without a plan' is demonstrated in the following example:

Figure 5.3 shows a student's spending plan for the first two terms of the academic year. Money will come into the student's account at the

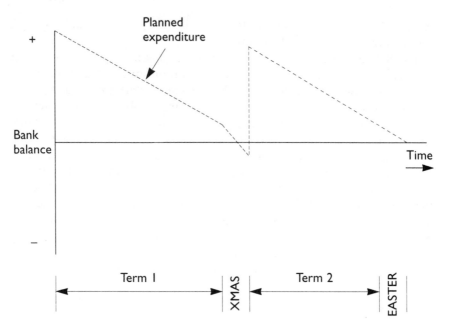

Fig. 5.3 Expenditure: initial plan

beginning of each term and the plan is an attempt to define at the outset a sensible level of spending. It might be thought that the plan is a realistic one, as it recognises that the account will be overdrawn at the end of the Christmas vacation, but, as shown in Fig. 5.4, the student was unable to stick to this spending plan. The line on Fig. 5.4 that is not on Fig. 5.3 shows how the student's spending actually varied over the first two terms. This shows that progress was indeed monitored, but does not show whether any control was exerted. The change in grade of the spending profile after approximately three weeks of the first term does, however, suggest that the student recognised that spending was out of control (by comparing actual outcomes with the plan) and made a decision to spend less – a controlling action. Unfortunately, although the rate of spending was brought roughly back to the original planned rate, the overspend in the first three weeks was never recovered. This might have been achieved had a revised plan been drawn up at the three-week point and a reduced level of spending accepted.

Plans are only attempts to lay out a course of action for the future and will frequently need to be adjusted. As the activities covered by the plan are actually carried out, the new information about how well we performed in those activities must be recorded at sensible intervals and the plan updated as necessary. This may prove that we are no longer likely to meet (say) our target completion date, and if it is all-important that the work is not delayed, it may be necessary to complete certain future activities at a faster rate than was originally envisaged. Such a decision involves a revision to the original plan, and this process of regularly checking progress against the current version of the plan and adjusting as necessary is typical of the control process necessary for effective project management.

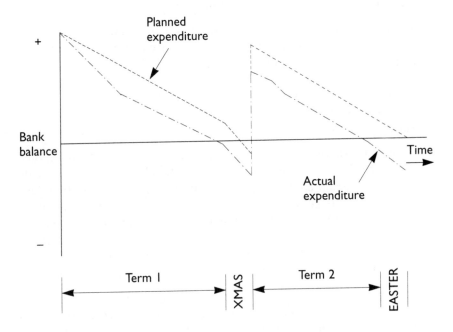

Fig. 5.4 Expenditure: what actually happened

Obstacles to effective planning

All of the parties traditionally involved in construction must plan to make sure that they fulfil their part in the effective construction of the project. The Promoter will need to ensure that s/he has sufficient funds to pay the Engineer and the Contractor. The Engineer must plan for a level of design resources to complete all of the design projects undertaken within their specified time limits. The Contractor must plan to:

- allow his/her estimating department to develop construction methods which when costed will enable a sensible price to be offered for carrying out the work and ensure a reasonable profit (*tender stage*);

- efficiently direct the various resources employed on the project, monitor and control progress and evaluate the effects of changes (*construction stage*).

It is the plans produced by the Contractor that will concern us from now on and, for the moment, the particular difficulties that make this process so problematical. These can generally be described as follows:

1. The fact that all projects are unique means that nothing is ever done in exactly the same way twice. We may have laid drains before, but never with this same crew or in this particular site location. Records of similar work in the past can therefore only ever be a guide to how we may be able to perform in the future.

2. In the UK, in particular, the weather has a considerable influence on the progress of construction work. While very cold weather will prevent concreting, it is cold and wet weather that will inhibit earthworks, and work needing the use of a crane may be affected by high winds. Predicting how the weather during a contract will affect the various activities comprising the contract is thus particularly troublesome, and yet the Contractor must allow for such delays as can be anticipated.

3. Ground conditions are notoriously difficult to predict and, even though extensive site surveys may have been undertaken, problems with the materials encountered in foundations frequently arise and often delay the works. It is generally recognised that once the job is out of the ground, the worst is usually past, but all construction projects will need to take support from the ground in some way. Although these delays need not be planned for by the Contractor, who can usually rely on the design being correct, they will affect the progress s/he can make.

4. Most construction projects will employ locally available labour, at least as part of the workforce. This means that some of the contract work will be carried out by employees who will not be committed to the project and will actually be out of work when the contract is complete. The effect that this will have on productivity is not easy to predict.

For the reasons given above, it is often recommended that project plans should not be produced in minute detail, but should be comprehensive without being overly prescriptive. This will affect the number of activities we have in our plan, which should be kept to the minimum needed to properly describe the work and allow control to be exercised.

Project plans

The complexity of projects can cause the plans used to describe them to become unwieldy and unintelligible. This is a danger, for if we cannot easily understand our plans, they are unlikely to be able to help us to control the work effectively. Simpler plans will typically be better presented and show fewer activities, but clearly will miss out some of the detail that we know to exist in the project. Getting the balance right between accessibility and complexity requires good judgement, but can be aided by using a master plan for the project, with further detailed plans for particular parts of the work. When the final plan is produced, it should help in a number of ways:

1. It should help predict the overall project time.

2. It should help to identify the activities which are critical to the achievement of that project time and allow us to understand how the project duration might be shortened.

3. By following the plan, or the revised plan, the Contractor will be able to direct the work in the knowledge that overall objectives have been considered.

4. Through knowledge of when the various activities should take place, the lead times for scheduling of subcontracts and material deliveries should be readily identified, allowing orders to be given in good time.

5. The plan should allow the Contractor to understand the level of resources needed to complete the work and allow the most efficient scheduling of those resources.

6. The Contractor should be able to use the plan to monitor progress and recognise when controlling actions will be needed to achieve essential objectives.

7. For delays to the works, the plan should permit the assessment of liability for any cost and time overruns.

Because the meaning of all of these items may not be completely clear at the moment, it is recommended that they be revisited when the rest of this section has been covered, when it is hoped that no mysteries will remain.

It is now time to learn about the planning techniques available in project management, and four techniques will be considered here. They are:

- bar charts

- the critical path method

- line of balance

- PERT.

Bar charts

The bar chart, sometimes known as the Gantt chart after Henry Lawrence Gantt, who is credited with having invented it, is probably the simplest and easiest to understand of all planning aids. To develop a bar chart for a project, the following steps must be followed:

1. Break down the project to be planned into separate, identifiable activities.

2. For each activity assess the likely duration.

3. Make decisions about when each activity will take place and represent

the activities on a chart with one row allocated to each activity and the timing and duration of each activity shown as a band in the *x*-direction.

Figure 5.5 shows an example of a bar chart for the construction of a reinforced concrete base, which has been broken down into five separate activities. Of course, the construction of such a base is a trivial exercise not worthy of being called a project, but it is quite adequate to explain the development and use of bar charts.

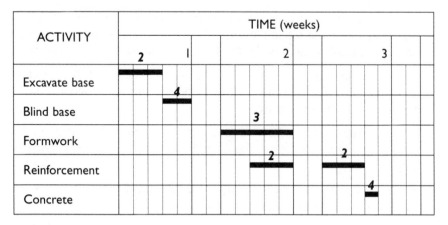

Fig. 5.5 Bar chart

Notes:

1. In general, an activity is a task which takes time to complete. It will usually also involve the use of resources, viz. labour, plant, materials.

2. A decision must be taken on the level of activity to be used in a plan, and whereas for this project the work has been broken down to the level of 'excavate base, blind base, etc.', it is easy to see that on a bigger project the construction of the base might be one activity. On an even larger project, a number of bases might be combined into an activity, 'foundations'.

3. To define a duration for an activity, we need to suggest a level of resources that will be used on that activity and, from past experience, predict the productivity of those resources. Knowing the amount of work in our activity, we can then estimate its duration.

4. Decisions about when each activity will take place in relation to the start of the project and to other activities will be made using knowledge of normal construction practice. For instance, it is evident from the plan that the base cannot be blinded (covered with low-quality concrete) until excavation has taken place. For a very long base, this might not be true as blinding could follow a few days behind the excavation, but that is not possible here. Also, the formwork must start before any

reinforcement is fixed, but once some of the formwork has been fixed, the steelfixing can then continue at the same time. Again this decision relies on an understanding of how we normally carry out such work.

5. The bar chart can be used as a control tool, as demonstrated in Fig. 5.6. After two weeks, the Contractor assesses progress on the project and finds that, although some activities were done as planned, others have taken longer to complete. From a knowledge of progress so far, the Contractor can predict that if work continues at the original rate, the job will be completed two days late. If this is unacceptable, something must be done, and we can see that by working the weekend after the second week, the lost time could be recovered.

6. The bar chart can be used to show the demand on resources, as shown in Fig. 5.7. The histogram below the bar chart shows the project's demand for labour throughout its life. The numbers above the activity bars represent the labour needed for each day that activity is under way, and when these are summed, the combined effect is shown on the histogram. This information can be used by the Contractor to plan recruitment or, if the demand is shown to be higher than the levels of labour that can be provided, the Contractor may have to adjust the scheduling of some of the activities. In some circumstances, this may delay the project completion date.

7. Although the bar chart has clearly been produced with a knowledge of the sensible and logical ordering of activities, this knowledge has not been incorporated within the project model. This means that if (say)

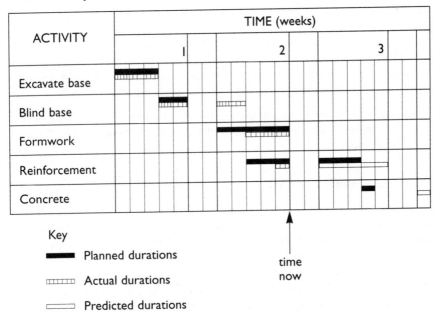

Fig. 5.6 Bar chart used for control

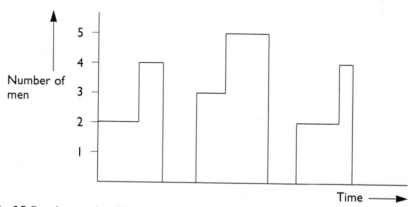

Fig. 5.7 Bar chart used to illustrate resource requirement

'excavate base' were to be delayed by one week, the plan does not, of itself, require that subsequent activities also be delayed (even though we know this must be the case). The technique known as CPM, the critical path method, will, however, enforce such an adjustment.

8. Because bar charts are very easy to understand, being a very visual depiction of a schedule, they are often used to show the results of a CPM model (*see later*).

Critical path method (CPM)

As suggested in the previous section, CPM is a technique that is similar in some ways to the bar chart method. It begins by breaking the project down into activities and allocating durations, as in the bar chart approach, but the essential difference is that in CPM we then take the process one stage further. This important additional step involves the building into our plan of the logical relationships we know to exist between the various activities. It uses the understanding we needed to schedule the activities in the bar chart and builds this understanding into the plan itself. This means we need a system that will allow us to model the fact that some activities *must* precede others, while some can proceed simultaneously.

Several terms are used to describe this general approach to project planning, and as the terms are not universally standardised, the meanings that will be used in this text will now be defined:

Critical path method (CPM)
This is the general term used to describe the method of planning that breaks down the project into activities and builds up a network that incorporates the relationships between the various activities. It is a deterministic model of the project, meaning that activity durations are given a single value (compare with PERT later). Other equivalent terms are *critical path analysis* (CPA) and *network analysis*.

Activity on node
A particular form of CPM, which portrays activities as nodes (rectangles) and the relationships between activities as links between the nodes. This system is also known as *precedence networks*, and will be the main system used in this text.

Activity on arrow
An alternative form of CPM in which the activities are shown as arrows and the junctions between successive activities are shown by circles and known as events.

PERT
PERT is the acronym for Programme Evaluation and Review Technique. This technique constructs a model of the project as in CPM, but in this approach activity durations are represented by a frequency distributio rather than by single-figure values. In this way, the uncertainty that is felt about the achievement of activities is incorporated into the project model. It is a probabilistic model whereas CPM is a deterministic model.

Activity on node and *activity on arrow* are alternative forms of CPM, whereas PERT is an alternative technique to CPM which will be considered in detail later in the chapter.

Activity on node

A CPM network is simply a combination of nodes (activities) and links between nodes, and although there are different types of links, these are the only symbols we need to understand. Different authors suggest different shapes for the nodes, but here the shape adopted will be as shown in Fig. 5.8.

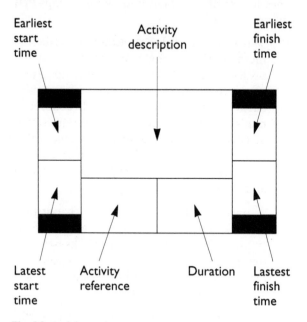

Earliest start time

Activity description

Earliest finish time

Latest start time

Activity reference

Duration

Lastest finish time

Fig. 5.8 Activity node

The meaning of the information which is recorded in each of the compartments will be explained later, but for the moment it is the different kinds of links we will consider. The most commonly used links are shown below:

Link

Explanation

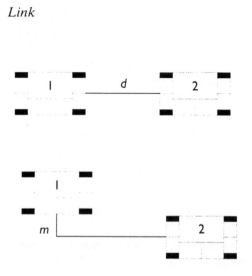

This is known as a *direct link*, and means that activity 2 cannot start until activity 1 is complete. If d is given a value, (say) 3, this means that there must be a delay of 3 time units between the end of activity 1 and the start of activity 2.

This is known as a *lead link* and means that m time units of activity 1 must be complete before activity 2 can begin. This kind of link is an easy way of showing an overlap between activities and can be useful where a second activity can only start when some of the first activity is complete.

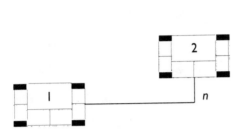

This is known as a *lag link* and means that when activity 1 is finished there must still be n time units of activity 2 to be completed. Where there are overlapping activities coming to an end, this is a common situation we will want to model.

Note that CPM diagrams always progress from left to right!

The direct link is the most frequently used. For example, in building a house the foundations must be constructed before the walls can be built and thus there would be a direct link between the end of the activity 'construct foundations' and the beginning of the activity 'build walls'. Such a link is often known as a *hard link* because the dependency must always hold. In other circumstances, direct links are used because the resources being used on one activity are needed for another activity. For example, where a contractor knows that there will only be one excavator available on the site, the different areas of excavation, which may be represented as different activities, will be shown as a series of activities connected by direct links. Such a link is often known as a *soft* link because the dependency is enforced only by resource limitations. If there were more excavators, some of these links would not be necessary. In general, it is recommended that soft links only be used where it is certain that the limitation on resources they represent is sure to hold when the project is constructed.

Lead and lag links are sometimes combined as shown in Fig. 5.9. Here the activities for construction of a long length of retaining wall are shown. Because it is a long length, it will be constructed in bays, and the lead link between 'excavate' and 'construct base' shows that once the first bay is excavated (after r days), the base for that bay can commence. Excavation will of course continue until all bays have been excavated, at which time the last bay will still need its base constructed, hence the lag of p days. In the same way, construction of the wall must follow behind construction of the base.

Having defined the symbols we need, the best way to proceed is to identify a simple project and to develop a CPM plan for that project. The project to be planned is shown in Fig. 5.10 and involves the construction of a single-span bridge.

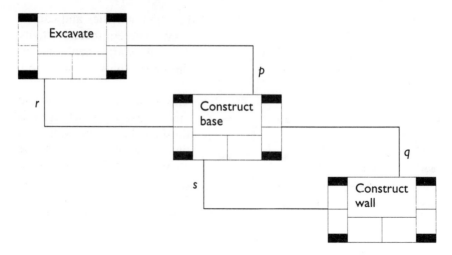

Fig. 5.9 Ladder activities

The embankments have already been placed under a separate contract and so need not be included in our plan. All that the project plan must include is the construction of two abutments and the bridge deck, backfilling both abutments, laying kerbs over the deck and surfacing the road over the bridge. The activities to be used in the plan are also shown in Fig. 5.10, where it can be seen that the abutments have been broken down into bases, stems and wingwalls. The North and South bases are to be shown as one activity. The only resource constraint involves the formwork for the abutment stems and wingwalls. As the North and South abutments are identical, only one set of formwork is to be provided for the stems and one set of formwork for wingwalls.

The network representing the construction of this project must now be developed and must incorporate the logic we know must hold, together with the resource constraints just detailed. It may take a number of attempts to develop an acceptable plan which properly depicts this understanding but eventually an acceptable network should result. This process is iterative, meaning that we will typically have to start by making a flawed attempt to construct the network and learn from the process, so that the next attempt is an improvement on the first.

The process, particularly on a complex project, may need to be repeated a few times before an acceptable solution is achieved. Figure 5.11 is a solution that fulfils all of the requirements.

Note the way the activities have been numbered. Later activities always have higher numbers than earlier ones.

To confirm exactly how this network models the construction of the bridge, the relationships shown will now be discussed:

1. **N & S bases**, although shown as one activity, clearly covers two areas

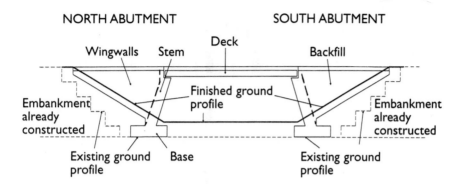

NORTH ABUTMENT   SOUTH ABUTMENT

Activity name	Activity reference	Activity duration (days)
North and South bases	1	12 14
North abutment stem	2	15 8
South abutment stem	3	15 8
North wingwalls	4	12 7
South wingwalls	5	12 7
North backfill	8	6 2
South backfill	7	6 2
Deck	6	20 12
Kerbs	9	2 1
Surfacing	10	2 1

Fig. 5.10 The project to be planned

of work and the North base is to be constructed first. When that is complete, after 6 days, the **North stem** can begin, and when the **South base** is complete, after 12 days, the South stem can be constructed.

2. The **N stem** must precede the **S stem**, as they both use the same form-work, and both stems must be completed before work on the **Deck** can begin.

3. Wingwalls cannot be constructed until the respective stem is constructed, therefore **N wwalls** follows **N stem** and must precede **S wwalls**, which need the formwork from the North wingwalls.

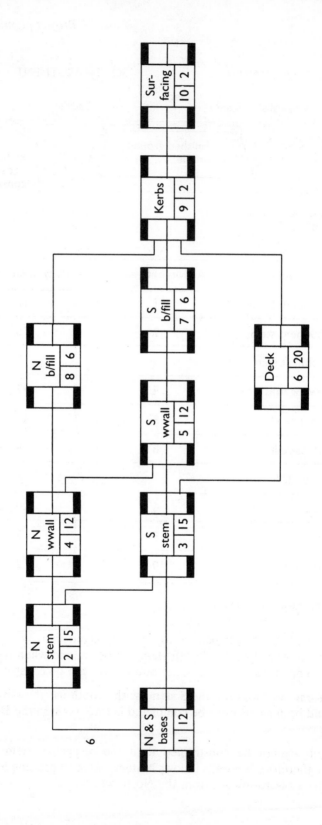

Fig. 5.11 The basic project plan

4. No backfill can be placed until the wingwalls have been constructed, and this is enforced by links 4–8 and 5–7.

5. The kerbs, which are laid both on the deck and off it, cannot be constructed until the **Deck** and **N b/fill** and **S b/fill** are all complete, hence the need for links 8–9, 7–9 and 6–9.

6. **Surfacing** cannot be laid until **Kerbs** are in place.

The next step is to use the individual activity durations to determine the minimum time in which the project can be completed. We need to use some of the boxes on the activity nodes to do this and will begin by working out the earliest start times and earliest finish times for each activity. Figure 5.12 shows the network with these boxes complete for all activities, and we can see that the earliest time at which the project can be completed is 60 days. To deduce this information, we must carry out what is known as the *forward pass*. We start with the first activity in the network and give it an earliest start time of 0, and then move forward through the network to determine earliest start and earliest finish times. Activity 1 has a duration of 12, and thus its earliest finish time is 12 days after it started, i.e. time 12. Because of the lead link, activity 2 cannot start until 6 days after activity 1 started, that is, at time 6, and will be completed 15 days later at time 21.

Whenever there is only one activity preceding the activity we are looking at, the process is simple; for example, activity 8 can only start when activity 4 is complete. The earliest finish of activity 4 is 33, and therefore the earliest start of activity 8 is also 33.

If we look at activity 9, there are three activities which must be completed before this activity can commence. Activity 8 has an earliest finish of 39, activity 7 an earliest finish of 54 and activity 6 an earliest finish of 56. Because activity 9 cannot start until all three activities are completed (that is what is meant by our logic diagram), the earliest start of activity 9 must be 56. We proceed in this way until all earliest start and finish times have been determined.

Next we must complete the *backward pass* through the network to determine the latest start times and latest finish times, and we begin by entering '60' as the latest finish time of the last activity in the network. This is done because we want to complete the project in minimum time and we therefore want the latest finish time to be equal to the earliest finish time. The results of the backward pass are shown in Fig. 5.13.

Working backwards through the network, we deduce the latest times that each activity can start and finish without delaying the whole project. Thus, the latest start time of activity 10 is 58. If it were any later, the minimum project completion time of 60 would be jeopardised.

In a similar manner to the forward pass, whenever there is only one activity following the activity we are looking at, the process is simple. Consider activity 9; the latest start of activity 10 is 58 and therefore the

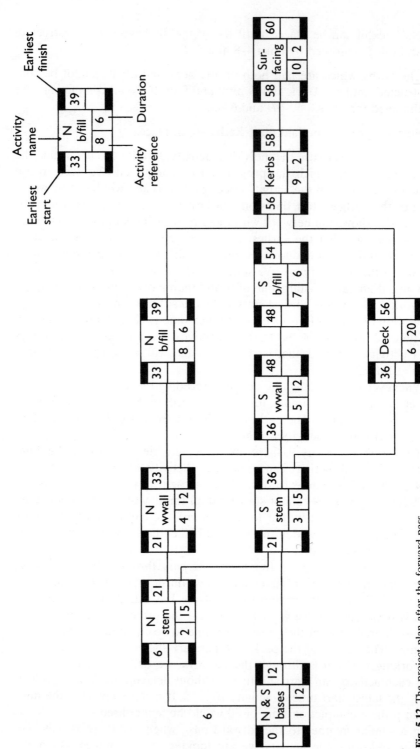

Fig. 5.12 The project plan after the forward pass

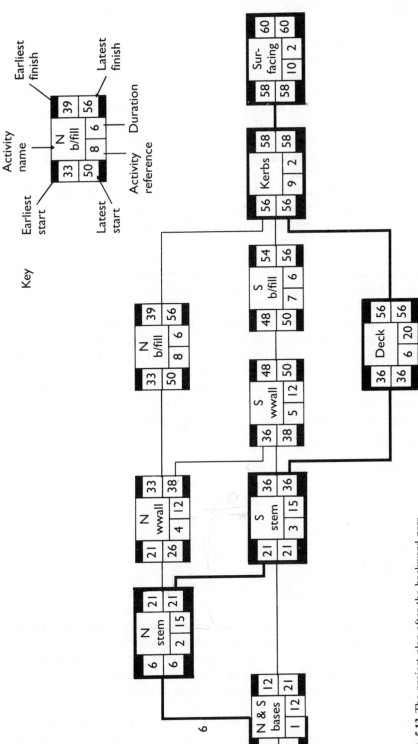

Fig. 5.13 The project plan after the backward pass

latest finish of activity 9 is also 58; any later and it would cause delay to the completion of the project.

When there is more than one activity following, we need to take more care. If we look at activity 3, this is followed by activity 5, with a latest start of 38, and activity 6, with a latest start of 36. The latest finish of activity 3 must be 36 and not 38. If we made the latest finish time of activity 3 equal to 38, the whole project would be delayed by 2 days. We proceed in this way, moving backward through the network, until the latest start and finish times of all activities have been calculated.

Having finished this part of the procedure, we now have a good understanding of when the various activities in the project must take place if the project is to be completed in the minimum time. If we look closely at this network, Fig. 5.13, we can see that certain activities have the same time for earliest start and latest start, and the same time for earliest finish and latest finish, and the difference between their start and finish times is exactly equal to their durations. These activities are called *critical activities* and have no flexibility in when they must be scheduled. They cannot start earlier than their early start/late start time because of the activities that must precede them and they cannot finish later than their early finish/late finish time, or the whole project would be delayed. All CPM networks that show completion in minimum time will always have at least one path through them of critical activities and this is known as the *critical path*. This path is shown on Fig. 5.13 by activities and links with a bolder outline. In this network, the critical path includes:

- the first part of **N & S bases**
- **N stem**
- **S stem**
- **Deck**
- **Kerbs**
- **Surfacing**

This is the longest path through the network, and if we sum the durations of these activities we will see that they add up to 60. Note that the second part of **N & S bases** is not critical. That could be completed as late as day 21 and still not cause the whole project to be delayed.

All of the activities that are not included in this list are non-critical and will have some flexibility in when they can be scheduled. This flexibility is called *float* and there are two main measures of float: total float and free float.

The *total float* of an activity is the maximum amount of time by which that activity can be delayed without delaying the whole project. If we consider activity 4, **N wwall**, the earliest it can start is day 21, and the latest it can finish is day 38. It has a duration of 12 and so its total float is 38 -

21 - 12 = 5 days. This is a general condition, and we can say that the total float of any activity can be calculated from:

(latest finish time of the activity) - (earliest start time of the activity) - (activity duration)

The *free float* of an activity is the maximum amount of time by which that activity can be delayed without delaying any subsequent activities which are starting at their earliest start times. Thus, for activity 4, which is followed by activities 8 and 5, the free float is 0 because although activity 5 cannot start before day 36, activity 8 can start as early as day 33. As this is the earliest that activity 4 can finish, it therefore has no free float. If the network was changed and activity 8 was removed, the free float of activity 4 would be 36 - 33 = 3 days. In general, free float can be calculated from the following expression:

(earliest of the early starts of following activities) - (earliest finish of the activity in question)

It should, however, be noted that this expression only holds for direct links between activities. When load or lag links are involved, the more general description of free float given earlier must be used.

In an attempt to make the understanding of the concept of float more tangible, a further diagram has been provided as Fig. 5.14 and is a different representation of the network shown in Fig. 5.13. It is the same type of diagram as is shown in the video (listed in the 'Recommendations for further study' at the end of this chapter), and requires some explanation before we can make any use of it.

Figure 5.14 should be seen as a board with slots a–a, b–b and c–c, into which the activities have been mounted. Thus activities 2, 4 and 8 are mounted in slot a–a and can slide horizontally along it. Similarly, activities 1, 3, 7, 9 and 10 are mounted in slot b–b, and activity 6 is mounted in slot c–c. All of the activities are time-scaled and thus their length in the *x*-direction represents their duration. The links between activities are enforced in three ways:

1. by having activities in slots adjacent to one another; thus the direct link between activity 2 and 4 is enforced by having activity 2 to the left of activity 4 in slot a–a. No matter how late activity 2 is, activity 4 can never start until activity 2 is complete. This is a proper representation of a direct link.

2. by the use of long single arms connected to individual activities; thus the arm at the end of activity 3, which is connected only to activity 3, will always ensure that this activity must be completed before activity 6 can commence. (Note that the long arm attached to activity 1 is attached at a point half-way along the activity. This represents the lead link between activities 1 and 2.)

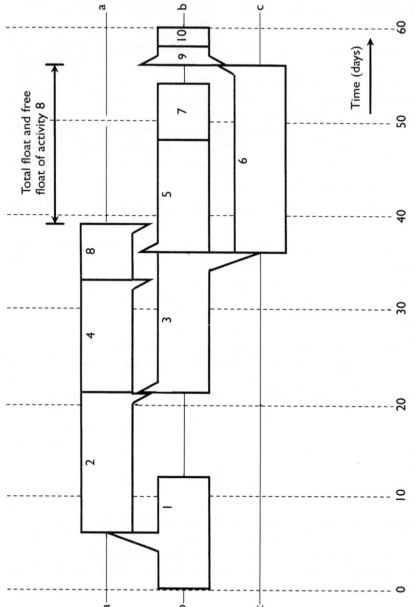

Fig. 5.14 The project plan, time-scaled, showing float of activity 8

3. by the use of shorter arms connected to two activities; thus the short arm connected to the end of activity 8 and the short arm connected to the start of activity 9 ensure that activity 9 will never be able to start until activity 8 is complete.

Note that all of the activities on this diagram have been shown starting at their earliest start times.

Having now described this diagram, we can use it to gain a better understanding of float. Remember that total float is the maximum amount of time an activity can be delayed without delaying the whole project, and consider activity 8 in Fig. 5.14. This activity is shown starting as early as possible and is followed only by activity 9, which is a critical activity and must start at day 56. Thus the maximum time by which activity 8 could be delayed without delaying the whole project is as shown in the figure, 17 days. Because activity 9 is critical, the latest time it can start is also the earliest time it can start, and as free float is the maximum time an activity can be delayed without delaying subsequent activities, the free float of activity 8 is also 17 days.

If we now consider activity 4 and look at Fig. 5.14, of the two activities following activity 4, activities 5 and 8, we see that activity 8 can start as soon as activity 4 is completed. This is true whatever time we choose to schedule activity 4. There is thus no time available to activity 4 without delaying a subsequent activity and thus the free float of this activity is 0 days. If we now look at Fig. 5.15, where activities 5 and 8 have been scheduled as late as possible, the maximum time-activity 4 could be delayed without delaying the whole project is shown on the diagram as 5 days. Thus the total float of activity 4 is 5 days. Note that if activity 4 uses all of its total float, it will use up all of the float of activities 5 and 7, and reduce the float available to activity 8. It is generally true that the total float of successive activities on a non-critical path is shared. If float is used by an earlier activity, there will be less available to following activities.

Activity on arrow

Although the preferred system for CPM is activity on node, it is important to be aware of the other available method, known as 'activity on arrow'. In this system, the symbols used are:

Symbol Meaning

⟶   activity – any task that consumes time, and will probably also consume resources

--→   activity with zero duration, used to define the logic

⊕   event – the start or completion of activities, which has no duration

As an example of a plan using this notation, the 'bridge' network has been produced in activity on arrow format and is shown as Fig. 5.16.

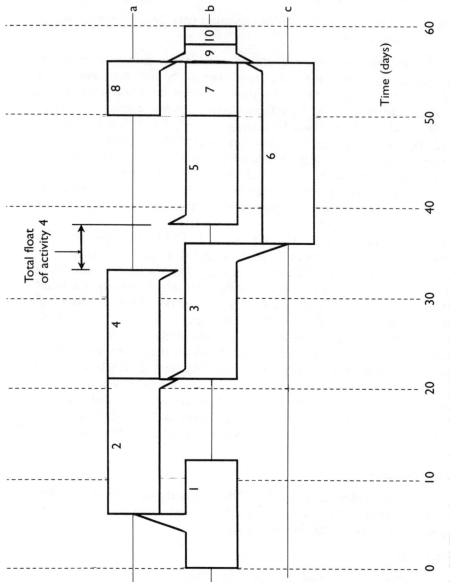

Fig. 5.15 The project plan, time-scaled, showing float of activity 4

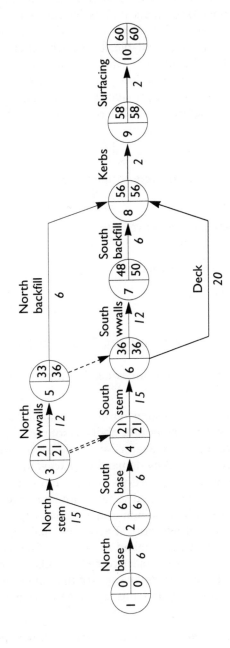

Fig. 5.16 The project plan: activity on arrow format

Resource scheduling

So far we have only considered time in developing the project plan, and no attempt has been made to check what level of resources will be needed to carry it out. Typical resources used in construction are plant (excavators, cranes, etc.), labour (steelfixers, joiners, labourers, etc.) and materials (steel, concrete, hard-core, etc.). If there is an upper limit (say) on the number of steelfixers we can make available, it is possible that our present plan will demand more than this number and thus the plan would not be feasible. We must therefore review our plan's requirement for the main resources to ensure that we can provide the levels that will be necessary.

Most activities will require a number of resources to be provided; for example, concreting is likely to need:

- a concrete gang, (say) 4 men   labour

- a crane to transport the concrete to the pour   plant

- a crane-driver   labour

- a skip to hold the concrete between the   plant
  mixer truck and the pour

- mixer trucks to deliver the concrete   plant

- mixer truck drivers   labour

- concrete   materials

and if these resources are not ready when the activity is due to start, some delay is likely.

When looking at the total resources needed for the project, there are two main objectives we may have:

1. *to limit the amount of a particular resource demanded by the project*: as described above. We normally talk about resource ceilings and if, for example, there are only 5 excavators available for a particular project, this is the resource ceiling. Any demand for excavators above this level cannot be met and if the plan does require more, the plan must be changed. Now the plan developed so far is not complete, as no decision has been made about when the non-critical activities should take place. By careful scheduling of non-critical activities that require excavators, we may be able to limit the demand to no more than is available. If it is not possible to achieve the objective by scheduling non-critical activities, we will typically have to delay some of the critical activities to stay below the resource ceiling and this will mean extending the project duration.

2. *to maintain a steady demand for a particular resource*: the ideal labour profile for a project is one that builds up steadily, remains fairly constant and then gradually reduces until the project is completed, as

shown in full lines on Fig. 5.17. The dotted labour profile on this diagram shows a very uneven labour demand, which would mean either that the workforce was not being used efficiently or that there would be regular hiring and firing. Neither is satisfactory, and so we can see that for some resources a relatively smooth profile is a good objective to aim for.

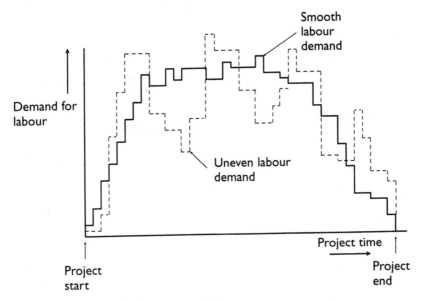

Fig. 5.17 Resource histogram

An example of resource scheduling will now be given, and at the same time the presentation of the CPM network as a bar chart will be introduced. Figure 5.18 shows the 'bridge' network in bar chart form with all of the activities starting at their earliest start times and the total float of non-critical activities indicated by dotted lines.

Most of the activities require labour throughout their execution, and the numbers required are given in Table 5.1. The aim is to achieve a schedule of activities that will give a smooth labour demand profile and will also keep the maximum labour required as low as possible. If we allow all activities to start at their earliest start times, the demand for labour will be as shown in Fig. 5.19. This shows a maximum demand of 12 men, and the profile is quite uneven. The first step in developing the required schedule is to show on the histogram the labour demand for all critical activities: there is no choice in when these activities must take place. Then, by moving non-critical activities within their available floats, we look for a schedule of non-critical activities that will achieve our objectives. Figure 5.20 represents such a schedule. It has a maximum labour demand of 10 and a somewhat smoother profile.

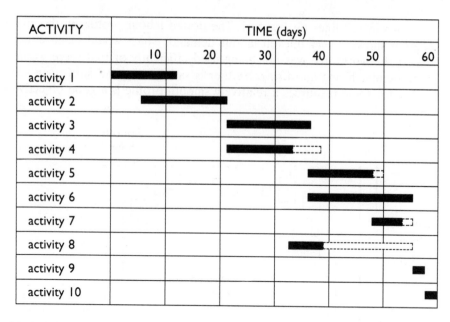

Fig. 5.18 The project plan shown as a bar chart

Notes:

1. The method used to identify the optimum schedule here is obviously trial and error, and it is easy to see that with a much more complicated project, such an approach would soon become unwieldy.

2. When using this manual approach, care must be taken to ensure that the schedule adopted does not defy the logic of the network, remembering that total float is shared between successive activities on a non-critical path.

3. The example is also rather artificial since no attempt has been made to discriminate between the different types of labour, but it is hoped that the simplicity allows the procedure involved to be better understood.

Table 5.1 Labour requirements of project activities

Activity	Labour required (no.)
1	4
2	5
3	5
4	4
5	4
6	6
7	2
8	2
9	1
10	0

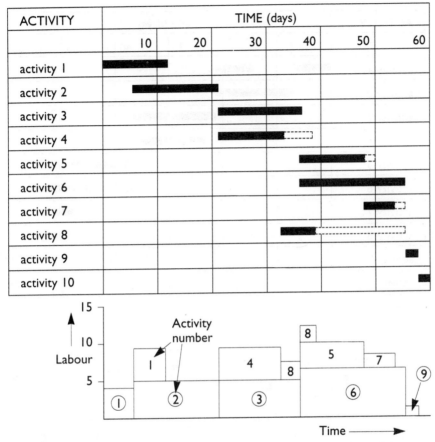

Fig. 5.19 Early-start histogram. Resource blocks with circled activity numbers relate to critical activities

4. It has been assumed that activity 1 can be split into two parts, and that the second part can be scheduled at its latest possible time.

5. The schedule shown in Fig. 5.20 shows the backfill of both abutments occurring simultaneously. No constraint on earthmoving plant has been stated, but it must be clear that a schedule that fulfils objectives for one resource may be unacceptable for another.

6. CPM software will allow all of the resource demands on a project to be held in memory and will permit the user to generate a schedule for the activities and then investigate the resource profile for each of the resources used. In this way we can easily see the effect of a particular schedule on all of the resources.

7. Activity 10, surfacing, is shown as not requiring labour. It is assumed that this work is subcontracted (carried out by another contractor) and therefore the labour required is not to be provided by the main contractor.

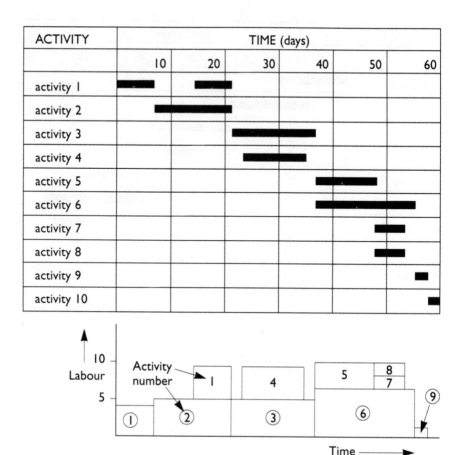

Fig. 5.20 'Smooth' histogram. Resource blocks with circled activity numbers relate to critical activities

8. One of the effects of resource scheduling is that non-critical activities will now have specific dates when they should be carried out. Although this scheduling is beneficial, because it allows us to use resources sensibly, it does remove some of the flexibility in the plan.

Review of networks

The point has already been made that project plans will need to be updated if they are to continue to be useful throughout the project. A plan that has not been updated on a substantial project will soon bear little resemblance to the work that has still to be carried out and will thus be ineffective. We therefore need to assess regularly the work that has been done and to adjust our plan accordingly: this process is known as reviewing the network. Once completed, the updated plan should tell us where we stand at the point of review and, of course, we may need to act if (say) the updated plan indicates that the job will be completed later than is acceptable. The agreed action should be recorded in the form of a revision to the plan.

With a CPM network, the review process is as follows:

1. At the point of review, all activities that have been completed should be given zero duration and any lead or lag that has already taken effect should also be given zero duration.

2. Activities that are under way at the point of review should be assessed to decide how much time they still need in order to be completed and this remaining time should become their new durations.

3. Activities that have not yet started should be considered to decide whether the original estimate of their durations is still thought achievable, given any new understanding that has been gained from work on other activities. If so, they will retain their original durations, but otherwise their durations should be amended to what is now thought more likely.

4. The logical relationships of the outstanding activities should be considered to decide whether they are still valid. If not, the network must be adjusted to reflect the new understanding.

5. By making the early start time of the first activity in the network equal to the time at the point of review and then carrying out a new forward and backward pass through the network, the updated plan may be produced.

6. Note that unless the project was exactly on schedule, there is a possibility that the critical path may have changed.

Using the 'bridge' network, an example of the review process will now be given. The contractor for the project has decided that reviews of progress should be carried out every 20 working days, and after the first 20 days the review showed that no changes to the plan were necessary. At day 40, a second review was carried out and the status of all activities was as shown in Table 5.2.

Table 5.2 Review data

Activity	Status	Remaining time to completion (days)
1	Completed	0
2	Completed	0
3	Completed	0
4	Completed	0
5	Under way	13
6	Under way	17
7	Not started	6
8	Under way	4
9	Not started	2
10	Not started	2

Activity 5, **South wingwalls**, is under way, and yet still has 13 days before it is completed. It should be noted that this activity was estimated as taking 12 days in the first instance, but from the experience gained in constructing the North wingwalls, the contractor has revised this estimate to 14 days, one of which has already passed. None of these matters need to be considered when carrying out the review procedure; it is only the remaining time to completion we need to know. Figure 5.21 shows the results of the review, following the procedure just described, and from this it is clear that the whole project is now delayed by 3 days. It can also be seen that the critical path has changed. Previously, the critical path went through activities 6, 9 and 10, but now it is activities 5, 7, 9 and 10 that are critical.

Figure 5.22 has also been prepared to show the results of the review. All of the activities that are complete have been removed, and those activities under way at the time of the review have been positioned so that their remaining time to completion lies to the right of the 'time now' line. By then positioning the other activities as early as possible, we see the same result as described above. (All of this procedure is equivalent to the procedure detailed above.) On this diagram it is easier to see what we must do if we wish to recover the 3 day overrun. If we accept that activities 9 and 10 cannot be completed in any less time, then it is activities 5 or 7 and 6 that must be speeded up. Although only 5 and 7 are on the current critical path, if we reduce the duration of (say) activity 5 by 3 days, the effect will only be to pull the project completion date back to 61 days. This is because after 2 days' shortening of activity 5, activity 6 will also become critical, and to then shorten the project any further will require that activities 5 and 6 are both shortened. If the contractor decides that he must recover the 3 days and that this is the way to do it, the plan should then be revised to show this. The change would thus be to reduce the duration of activity 5 by 3 days and of activity 6 by 1 day.

It should be noted that the term 'updating' has been used to describe the process of reviewing progress at intervals, and the updated network is the result of that process. Where the information thus generated causes the contractor to then amend the plan (say) to pull it back on schedule, the term 'revising' has been used, and the network that results is called the revised network. Clearly, where no action is needed the two will be the same, but it should be recognised that at each point of review, there may be an updated version of the plan and also a revised version.

Line of balance

Some projects consist of a set ordering of activities that must be repeated several times to produce a number of units of finished work. A good example of such a project from the building sector would be the construction of a number of houses on an estate. In civil engineering there is less obviously repetitive work, but the construction of a multi-span viaduct would

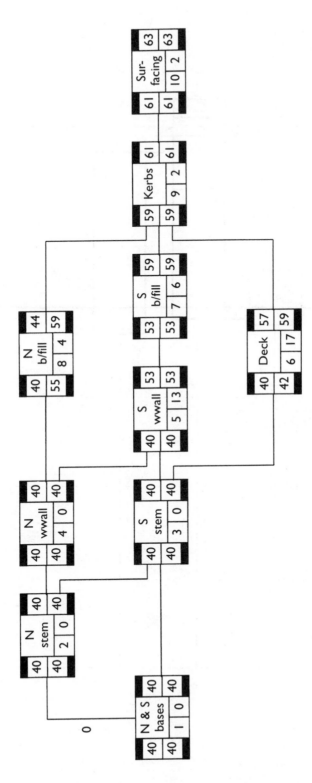

Fig. 5.21 Network plan at review date

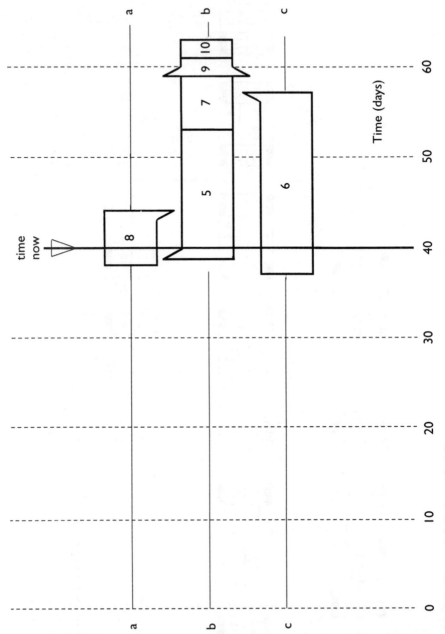

Fig. 5.22 Time-scaled plan at review date

fall into this category, with each span being represented by a sequence of activities which are repeated until the viaduct is complete. These projects could be planned using either bar charts or CPM, but there is an alternative technique devised specifically for such repetitive situations: the line of balance technique. A multi-span viaduct is used as a numerical example later in this chapter.

The work necessary to complete each unit is represented in the form of a series of activities, and the activities that would be needed for the construction of a number of mass concrete base units are shown in Fig. 5.23. For each base to be constructed, the site of the base must be excavated, the bottom of the base blinded, formwork erected and then concrete poured. Each activity in this sequence must be fully completed before the next activity can start.

Fig. 5.23 Repeated sequence of activities

This sequence of activities must be repeated until all required units are complete. Thus, if we are required to construct (say) two of these units per week, this can be shown graphically as in Fig. 5.24.

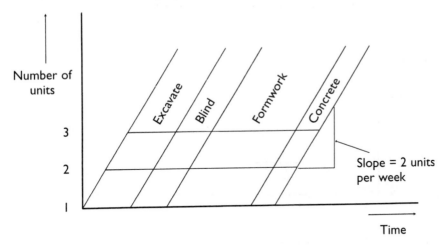

Fig. 5.24 Line of balance plan – simplest form

However, we must not forget that this is simply a plan, and if our plan is not realistic it will not give us reliable or useful information. As it stands, any deviation in the rate of working of any of the activities shown will immediately affect subsequent activities and hence the total project time. To allow for the variation in rates of working, which we know will occur, we allow a time buffer between each activity as shown in Fig. 5.25.

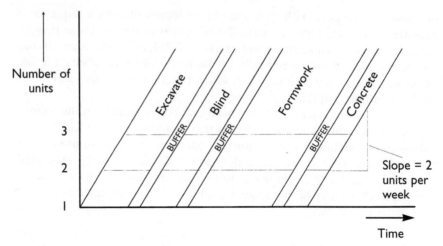

Fig. 5.25 Line of balance plan including buffers

There is one more alteration needed to increase the realism of our plan, and this requires some understanding of the natural rate at which activities take place most efficiently. For most activities, there is an optimum gang size, and with such a gang the activity in question will be completed within a certain time. Thus, if we only have one gang (A), this dictates the rate at which the activity can progress, as shown in Fig. 5.26.

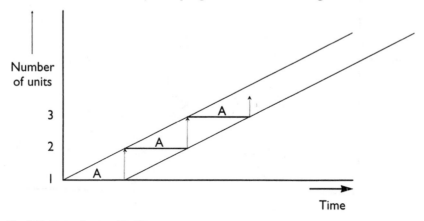

Fig. 5.26 Natural rate of build – one gang

We can, however, increase that rate by employing more than one gang. Figure 5.27 shows two gangs, A and B, working on this activity, which is now being completed twice as fast as shown in Fig. 5.26. Gang A starts on the first unit and then moves on to the third unit, while gang B starts on the second unit and moves on to the fourth unit, and so on. . .

Thus, if each activity is to be completed with maximum efficiency, it is unrealistic to expect them all to be carried out at the same rate, and the more realistic plan will look like Fig. 5.28.

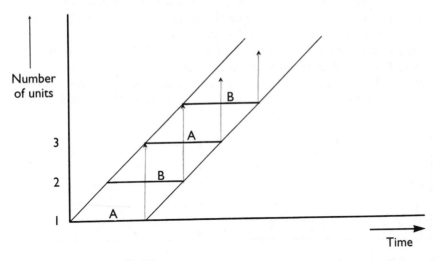

Fig. 5.27 Natural rate of build – two gangs

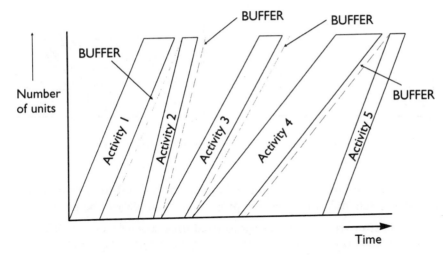

Fig. 5.28 Realistic line of balance plan

Notes:

1. If the objective is to produce completed units at a particular rate, it will presumably be acceptable to produce at a faster rate, and it will probably not matter that none of the activities are actually being completed at the required rate. Note that in Fig. 5.28, the rate of completion is dictated by the rate at which activity 5 is completed.

2. From an understanding of the resources being applied to each activity, a histogram of total resource requirement for a particular resource can be produced.

3. If the overall project time must be reduced, this can be done in different ways. One way is to speed up a particularly slow activity. In the case
shown in Fig. 5.29, by speeding up the completion of activity B, the project is completed earlier. Another way is to stop an activity which is
particularly fast and restart it at a later date (Fig. 5.30). Here, by stopping C and restarting later, the overall effect is to reduce total project
time. This is because the last activity can now be started earlier,
only needing to be the buffer distance away from the second part of
activity C.

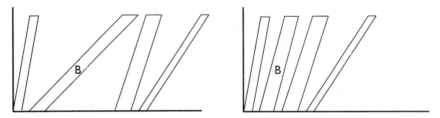

Fig. 5.29 Reducing project time by speeding up an activity

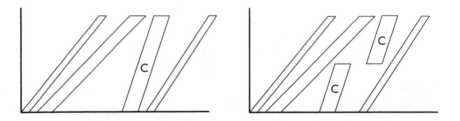

Fig. 5.30 Reducing project time by stopping and restarting an activity

To show the kind of calculations likely to be needed to develop a line
of balance schedule for a specific project, an example will be given:

A multi-span viaduct is to be constructed using a simply supported super-
structure, comprising precast beams with an in-situ concrete deck slab. The
design has been rationalised to provide standard height piers at regular intervals as shown in Fig. 5.31. The construction of the viaduct between abutments
is best considered as the repeated production of combined pier and deck sections, and the activities required to complete each section are also shown in
Fig. 5.31. The viaduct has 21 identical spans, and to stay on target it is estimated that these must be completed at a rate of one every week. Using the
activity data in Table 5.3, we must prepare a line of balance schedule for the
construction of the viaduct between abutments, assuming that each of the teams
involved works at its natural rate. The men will work an 8-hour day, 5 days a
week, and a minimum buffer of 3 days must be allowed between activities.

Note that the placing of beams for any deck will always be completed
within 1 day.

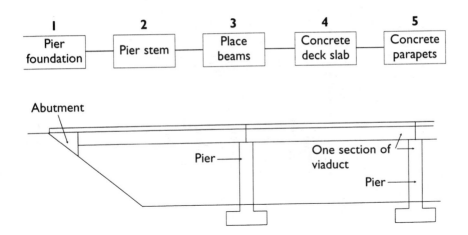

Fig. 5.31 The project to be planned

Table 5.3 Activity data for building a multi-span viaduct

Activity	Man-hours per unit	Optimum gang size
Construct pier foundations	160	4
Construct pier stem	200	4
Place beams	24	3
Concrete deck slab	180	3
Concrete parapets	100	2

The solution is obtained by generating Table 5.4, in which the figures in brackets in columns G and H are rationalised values. Columns A, B and D are given. Column C shows the gang size needed to produce one section of viaduct per week. Column E shows the actual number of men to be employed on each activity, which will always be a multiple of the figure in D. This involves a decision about how many gangs should be used to get closest to the required rate of build. There will thus be 2 gangs on each of activities 4 and 5. Column F shows the number of units constructed per week using P men (column E/column C). Column G shows the time for one gang to complete one operation – the width of the activity bands. Column H shows the distance along the *x*-axis from the start of the first activity to the start of the last (20 = 21 - 1).

Using the information in columns G and H and remembering to allow for buffers, the schedule can now be drawn as shown in Fig. 5.32. From this schedule, we can determine the numbers of each activity that must be completed by (say) day 80, to stay on target (15 units of activity 1 and 11 units of activity 2).

It is also clear that the overall project duration can be reduced considerably if it is accepted that the beam placing for each section of viaduct (activity 3) need not be completed on consecutive days. The slope of this activity would thus be made less steep, which would reduce project duration by as much as 55 days.

Table 5.4 Line of balance calculations

A (Activity)	B (Man-hours)	C (Theoretical gang size)	D (Men per operation)	E (Actual number of men employed)	F (Natural rate of build)	G (Time per operation)	H (Elapsed time between starting first and last operation)
	(M)	($1*M/40$)	(N)	(P)	(Q, (no./week))	($M/8*1/N$)	($20*5/Q$)
1	160	4.0	4	4	1.0	5	100
2	200	5.0	4	4	0.8	6.25 (6)	125
3	24	0.6	3	3	5.0	1.0	20
4	180	4.5	3	6	1.33	7.5 (8)	75
5	100	2.5	2	4	1.6	6.25 (6)	62.5 (63)

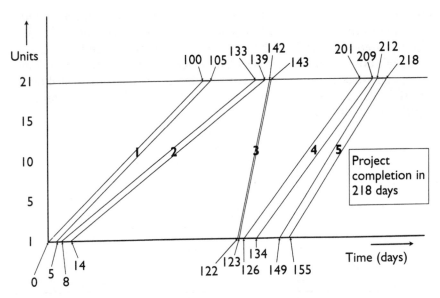

Fig. 5.32 Completed line of balance plan

PERT (programme evaluation and review technique)

Most projects in the construction industry are composed of activities whose durations we are likely to be able to forecast reasonably accurately. That is not to say that we would necessarily feel confident that the actual time taken to complete the activity will be equal to the estimated time, only that with our knowledge of the work in the activity and the productivity of the resources we intend to use, we feel able to specify a one-figure duration. There are some projects, however, in which the activities are so novel or so far removed from our experience that we feel very uneasy about giving such an estimate, and for such projects it may be that the use of a deterministic CPM plan would be inappropriate.

The planning technique PERT involves the construction of a network of activities in a similar fashion to CPM, but will also typically include 'project milestones'. These are important stages in the execution of the project, such as 'completion of earthworks', 'installation of mechanical plant completed', etc., which will usually be represented as activities with zero duration within the network. By then recognising a range of durations for each time-consuming activity and building this into the plan, one can produce a probabilistic model of the project. The uncertainty that is felt about activity durations is accommodated by giving three durations for each activity: a, an optimistic time that assumes all will go well; m, the most likely time – our best guess at a single-figure estimate; and b, a pessimistic time that assumes everything goes wrong.

The optimistic and pessimistic time are further defined as those times that we might improve on 1 time in 100 (optimistic) or which might turn out worse 1 time in 100 (pessimistic). A graphical representation of the situation we are trying to model is shown in Fig. 5.33, where the area under the curve to the left of *a* is 1/100th of the total area under the curve; likewise the area to the right of *b* is 1/100th of the total area under the curve. Although not reproducing exactly this shape, the beta distribution comes close, and since it can be adjusted to skew or symmetrical distributions and the calculations for its parameters are simple, it has been adopted to represent activity durations in PERT.

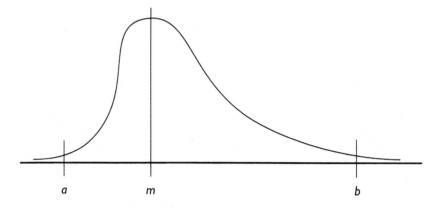

Fig. 5.33 Beta distribution – activity duration

The expected mean duration of this activity, adopting the beta distribution, is given by

$$t_e = (a + 4m + b)/6$$

and its standard deviation is given by

$$\sigma = (b - a)/6$$

Having calculated expected mean durations for each activity in the network, we can now do a forward and backward pass and determine the critical path using these durations. PERT then concentrates solely on this critical path, ignoring all others. The single value we now have for the project completion time is only the mean value, and, of course, as the activities contributing to that value are represented by frequency distributions, we would also expect the project completion date to be represented by such a distribution. The central limit theorem applied to this situation states that for a sufficiently large number of activities on the critical path, the distribution of the project completion time will be normal with a variance (σ^2) equal to the sum of the variances of all the activities on the critical path. Thus the mean project completion time is given by

$$M = \sum_{1}^{n} t_e$$

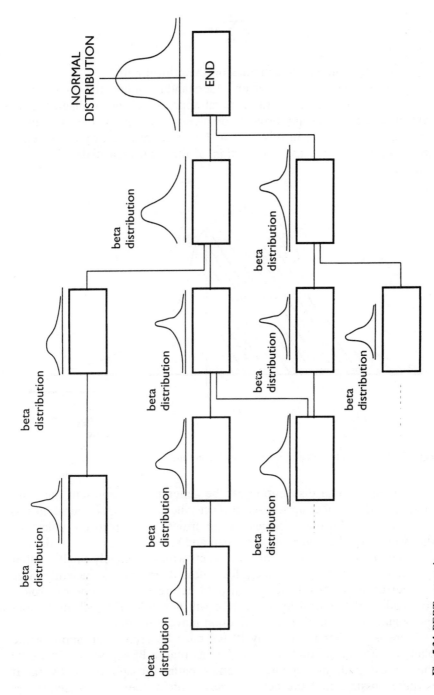

Fig. 5.34 PERT network

and the variance of project completion time is given by

$$\sigma^2_{\text{project}} = \sum_1^n \sigma^2_{\text{activity}}$$

where n is the number of activities on the critical path.

This means that we now have an understanding of how the project completion time may vary, and the general approach is shown diagrammatically in Fig. 5.34. For any project completion time, say x, the probability that we will complete in this time or less is represented by the hatched area under the curve as a proportion of the total area under the curve (Fig. 5.35).

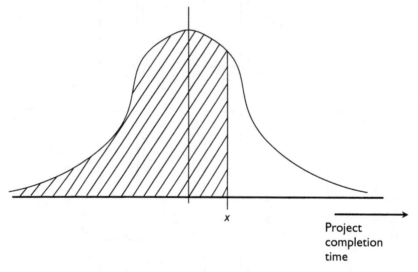

Fig. 5.35 Normal distribution of project completion time

A similar normal distribution can be generated to represent the completion of the project up to each project milestone and, of course, in this instance it will be the activities on the critical path up to the milestone that will be used to calculate the mean and standard deviation of the distribution. In such situations, or on projects with a relatively small number of activities, care must be taken, for unless there are a sufficiently large number of activities on the critical path before the milestone or project completion, the central limit theorem will not hold. This will mean that the assumption of a normal distribution may be invalid.

There is another reason why PERT must be seen as an approximate planning model, and that is due to the fact that in PERT we consider only the critical path and ignore all other paths through the network. It is quite possible that the network may contain one or more sequences of activities which are very close to being critical and which should sensibly be taken into account when deriving information about completion times.

It is certainly true that the principal planning techniques used in the construction industry are bar charts and CPM and that PERT is seldom, if ever, used. However, I believe that by recognising the additional sophistication of a probabilistic planning tool, a greater understanding of the practical limitations of CPM may result.

Recommendations for further study

Scott, S. (1993) *Dynamic Project Planning*, University of Newcastle upon Tyne, video – 35 minutes. This video is available from the University of Newcastle upon Tyne and was produced by the author to illustrate the CPM planning technique. In the video, a project plan is developed using the normal activity on node diagram and then using the dynamic time-scaled diagram.

6 The law of contract

Introduction

Most civil engineers will be involved with contracts throughout their working lives. These contracts may be for services, such as the design service offered by a consultant to a client; they may concern the provision of labour and materials essential to allow a contractor to proceed with the works; or they may relate to the construction of an entire project. In the course of their employment, engineers may be arranging contracts, possibly agreeing what work is involved and what price is to be paid. They may be administering contracts, which means that they will be making sure that work is carried out correctly, dealing with any difficulties and agreeing the level of payments for additional extras. Often, they will be responsible for work that is carried out as part of a contract. For example, the engineer working for a contractor must ensure that the quality of the contract work is up to the standards required by the contract. When there are disagreements about what work is covered by a contract or what payments should be made (and these occur frequently), it will be engineers, in the first instance, who must attempt to resolve these problems. Although clearly not trained in the law, it is nevertheless important that engineers have some understanding of how legal professionals operate and, in particular, of the law as it relates to contracts. Contract law is a complex area of study that develops over time, and thus the understanding that can be gained from a short introduction is clearly limited, but still worth having.

Perhaps the first point to recognise about the law of contract is that it relates to private arrangements and does not involve the police or the state. Individuals or organisations will enter into contracts with each other, and if they are not satisfied with the service they receive or the goods they acquire through the contract, they must make a claim for a remedy in the civil courts. The parties to the contract will typically have to pay for their own legal representation, which they may recover if they win the case. As will be seen in Chapter 7, which deals with the ICE *Conditions of Contract*, disputes on major construction contracts are unlikely to come to court, but rather will be dealt with in an alternative way: by conciliation or arbitration. Whether the dispute is heard in a court or in an arbitration hearing, the parties will still have to fund the proceedings. Sensible resolution of disputes between the parties, without involving outsiders, is thus always to be preferred. This is more likely to occur if engineers have some legal knowledge.

A contract is thus a legally binding agreement that gives rise to obligations that are enforced by law. That does not mean that the legal profession plays a part in every contract; the majority of contracts are executed with both parties satisfied with their involvement and these never come to court. However, when there is a dispute, provided that the courts are satisfied that a valid contract existed, they will enforce the details of the agreement. Over the years, the understanding of what constitutes a contract has been refined as a result of a variety of cases on which judgements have been made, and the courts will look for a number of features in any agreement before recognising it as a valid, legally binding contract. These features will now be considered in turn.

Offer and acceptance

For an agreement to have been reached it is expected that one of the parties will have made an offer which the other party has accepted. If they can identify the offer and the acceptance, the courts can confirm that such an agreement has been made. On a major construction contract, where tenders have been invited from a number of contractors, the offer will be each contractor's tender and the acceptance will usually be a letter from the client to the contractor that has submitted the most favourable proposal. In this circumstance, there is typically no difficulty in clearly recognising the offer and the acceptance. Even though there were a number of offers, only one will have been accepted and it is with the chosen contractor that the contract will exist. Not all situations are as transparent as this, and as a result of cases that have come before the courts, the meanings of 'offer' and 'acceptance' have been further defined. The offer must be a definite offer and the acceptance must be an unqualified acceptance by the other party. If the offer is accepted with conditions, then there has been no actual agreement between the parties (unless, of course, the first party later accepts the conditions stipulated by the second party). This problem of offer and counter-offer is demonstrated in the case *Butler Machine Tools Co. Ltd* v. *Ex-Cell-O Corporation*, which is represented diagrammatically in Fig. 6.1. Following an inquiry from the buyer, who wished to purchase a machine tool, the seller sent a quotation which had on the back standard conditions of sale that the machine tool company normally operated. These included a price variation clause which would allow the

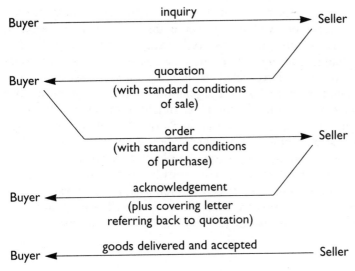

Fig. 6.1 Battle of the forms

seller to add to the quoted price to compensate for any increase in the cost of the goods incurred by the date of delivery. The buyer then sent an order for the machine, but on the back of the order was a set of conditions the company normally adopted for its purchases. These did not allow for any variation in price, and at the foot of the order there was an acknowledgement slip to be torn off and returned. This slip stated that the order was accepted on the terms and conditions on the back of the order. The seller returned the acknowledgement slip, duly signed, but with a letter stating that the order was being accepted in accordance with the original quotation. When the machine was delivered, the sellers claimed an additional £2,892 as a result of price increases, but the buyer refused to pay. The case went to court, where it was held that the variation-of-price clause was not a part of the agreement between the parties and thus no additional payments should be made. Because this kind of case results from the two parties trying to impose their form of agreement on the other, it is usually referred to as the 'battle of the forms'.

The findings of the court in this particular case are not what should concern us. What is important is to recognise the confusion that can be caused when the acceptance of an offer is qualified. Although this problem is unlikely to occur in the main contract for a construction project, in other contracts, notably for goods and in subcontracts, such difficulties have arisen.

Certainty

Where the courts cannot determine what the parties intended because they expressed themselves in such imprecise terms, and where these vague terms were essential to the agreement, the courts may decide that no contract exists. This does not mean that failure to define an important aspect of the agreement will always cause the contract to be seen as invalid. If the uncertainty can be resolved in such a way as to give effect to the contractual intention of the parties, the courts will do so. An example of lack of certainty is revealed in the case *Lind (Peter) and Co.* v. *Mersey Docks and Harbour Board.* The plaintiff, Peter Lind, submitted two tenders for the construction of a container freight terminal: one was fixed price, the other contained a price variation clause. The defendant accepted the 'tender' without making it clear which one was being accepted and no contract was signed. When prices rose, the plaintiff asked that the prices be revised but the defendant refused. In court it was held that no contract existed either on a fixed price or price variation basis as there was an obvious ambiguity that the courts were unable to resolve. The contractor was to be paid on a *quantum meruit* (as much as is deserved) basis for the work done.

Consideration

The courts will not enforce an agreement unless it contains an element of bargain. Where one party is making a gift of services or goods or money to another party, who gives nothing in return, there is clearly no bargain and therefore no contract. In these circumstances, if the gift were only promised and not actually given, the law would not enforce the promise. To test whether a bargain exists, the doctrine of consideration is used, and the question asked is, 'what consideration does the promisor [person making the promise] get in exchange for his/her promise?' Where there is no consideration, there is no bargain and therefore no contract. In construction, the consideration is usually obvious. The contractor pays the material supplier in consideration of the materials being provided, the client pays the contractor in consideration of the contractor constructing the works. If the parties have deliberated over their agreement in such a way as to fix the level of consideration, the courts will not interfere, even though it may appear a particularly bad bargain for one of the parties. Thus there must be consideration for a valid contract to exist, but the consideration need not appear reasonable to outsiders.

A good example of a promise which was not enforced owing to lack of consideration is reported by Abrahamson (*Engineering Law and the ICE Contracts*, 4th edition; *see* 'Recommendations for further reading' at the end of Chapter 7). A contractor engaged on the construction of a project refused to do a part of the work unless the employer agreed to pay for it as an extra. In fact, the work in question was included in the contract work which the contractor was bound to do for the original contract price. The employer said he would make extra payments, the contractor did the work and then the employer refused to make the extra payments promised. In court it was held that the contractor gave no consideration for the employer's promise to pay him extra and so the promise was not binding.

Not all contracts require consideration, and where a party does want to make a gratuitous gift and be legally bound by it, this can be done by making the promise in writing by deed. A deed must be witnessed and must make it clear on its face that it is a deed. Most contracts, however, are called 'simple' contracts. These are not deeds and do require consideration.

Capacity

A party's capacity to enter into a contract is unlikely to cause problems in construction. Adults of sound mind, companies, local authorities and statutory corporations are generally permitted to enter into a wide range of contracting arrangements, although there are limitations. A local authority can only enter into contracts in areas defined by statute and a private company is limited by its memorandum of association.

Persons under 18 years of age are very limited in the contracts they can make, and although not specially relevant, it is of interest to know that a contract entered into by a drunk may not be binding. If s/he did not understand what s/he was doing and the other party knew that to be the case, no contract will be enforced.

Intention to create legal relations

In commercial transactions, it will generally be presumed that parties who have made an agreement intend that the agreement be legally binding. If they do not intend this, they should say so explicitly. During the course of normal business in the construction industry, agreements between organisations or individuals will therefore be construed as intending to create a contract. However, social and domestic arrangements are not normally entered into with such a purpose. A variety of casual agreements between friends and family members, some of which will involve money, come under this heading. The courts wish to prevent frivolous claims from such agreements having to be dealt with in a legal arena and believe that it is better that these disputes be settled between the parties themselves, informally. As already intimated, however, such casual agreements are unlikely to create problems in the construction industry.

Breach of contract

There are instances when performance of the contract becomes impossible – for example, owing to outbreak of war – and in these circumstances, the contract is said to be frustrated. Where this occurs, the contract may be ended and a settlement should then be made. However, there are often circumstances in which one of the parties fails to honour his/her agreement and this is known as a breach of contract. When a breach occurs, the party who suffers will typically seek damages from the other party, and these damages are invariably in the form of a sum of money intended to compensate for losses. Where such disputes arise on a construction contract, the parties should make every attempt to resolve the matter themselves. If they are unable to do so, the procedures of conciliation or arbitration, if written into the contract, may be instigated to arrive at an impartial judgement. Where no arbitration clause exists, and if one party feels sufficiently aggrieved, the case must be taken to court.

Legal interpretation of documents

One of the main documents likely to be examined to try to identify who is responsible for any dispute is the ICE *Conditions of Contract*, 6th edition (ICE6), the conditions that govern many construction contracts in the UK. Even a cursory reading of these conditions will show that the wording is not easily accessible to the lay person. Many of the clauses are complex and difficult to follow, often containing very long sentences with little punctuation. They are written like this in an attempt to make the meaning of the words unambiguous. When a dispute does arise, the parties will consult the contract documents, and in particular the *Conditions of Contract*, to find out what remedies, if any, are available. There are several clauses in ICE6 which identify how specific problems should be dealt with and who is responsible. Even when the actual problem which has occurred is not obviously addressed by the *Conditions*, the most relevant clause(s) will be selected and will be read as though those who drafted the clause had this special problem in mind. This approach is in line with the courts' general wish to uphold the actual agreement reached between the parties. Although many arguments about disputes will not take place in court or even in an arbitration hearing but will be resolved on site, it is still the legal approach to interpretation of documents that will typically be adopted. This means that it is not uncommon for engineers representing the contractor and the client to be arguing about the precise meaning of a phrase or sentence on which the outcome of a particular dispute rests.

Precedent

It was stated earlier that the law relating to contracts develops over time, and this is the result of new cases being heard and new judgements being made. In English law, cases are heard before a variety of courts whose standing ranges from the highest court, the House of Lords, to County Courts with the Court of Appeal and the High Court in between. In general, the principles for decisions adopted in one court are binding on lower courts, and thus the cases heard can be said to be not just examples of the law, but the law itself. Since arbitration became fashionable for dealing with contractual disputes, there has been little additional case law available and it is for this reason that the cases quoted in contract law are often quite dated.

Recommendations for further reading

Tillotson, J. (1995) *Contract Law in Perspective*, 3rd edition. Cavendish Publishing, London. 272 pp.

7 Conditions of contract

Introduction

What are conditions of contract?

Conditions of contract are a set of rules that govern the construction process for a project. Provided the analogy is not stretched too far, they can be likened to the rules for a complicated board game: Monopoly, for example. They identify who are the 'players', and what they must do to play the 'game' properly. Clearly, the business of construction is not a pastime and is also much more complex than the most intricate of board games. That said, however, the need for a set of regulations to ensure that 'players' know how to play the game has some parallels with the need for a set of rules to govern the construction process.

On a much more mundane level, the conditions will usually be a softback paper document, selected for a particular project by the Engineer working for the Promoter.

Why do we need them?

Our current understanding of the contract between the Contractor and the Promoter is that it is for a defined amount of work, specified in the drawings, to be carried out for a particular price. If that were really so, then any change that needed to be made to those details because (say) a mistake had been made in the design process would involve a renegotiation of the whole contract. In fact, the contract is much more flexible than that. It allows for changes to the original details to be made and recognises that additional payment and possibly additional time may need to be given to the Contractor. The powers to make such changes and the Contractor's duty to accept them are spelt out in the conditions of contract, as is the Contractor's right to extra payment and, where necessary, extra time. This is just one example of the kind of area that is addressed in this most important document. Many more will be revealed in later sections.

What do they contain?

As we will see later, a variety of different contract conditions are available for use on construction contracts, but they will all cover similar material. Some of the commonly occurring topics are as follows:

Duties and powers of the site representatives

Clearly the Contractor will have a presence on the site, but there will also be someone to represent the Promoter's interests, whose main function is to ensure that the job is constructed properly. Both the responsibilities and the powers of these two parties will be defined.

Subcontracting

Although the contract has been won by a particular contractor (known as the main contractor), that contractor may not do all of the work. Indeed, it is common for main contractors to enter into agreements with other contractors who will each do a part of the contract work. This is called subcontracting, and the main contractor may have several subcontractors. On one contract, the main contractor may have agreements with different subcontractors to carry out the piling, the earthworks and the drainage, while s/he oversees their work and completes the rest in-house. The conditions of contract will usually specify the extent to which subcontracting is permitted and any control that the Promoter wishes to have over the choice of subcontractors.

The contract documents

Only the contract documents should be consulted to determine the exact details of the agreement between the Contractor and the Promoter. Exactly what these documents are will be defined, and procedures for providing additional or updated documents may be stated. It is not uncommon for detailed drawings, which were not available at tender stage, to replace outline drawings on which the contractors tendered, at some time during the construction of the works.

Insurance

During construction, accidents can occur which cause damage to the workforce, to the general public or to parts of the works already built. There will usually be a section of the conditions of contract which specifies who should be responsible for any damage and may also require that insurance cover be obtained.

Materials and workmanship

The Contractor will invariably be responsible for ensuring that the quality of the materials and workmanship are as required by the contract and this, together with the procedures for testing and dealing with substandard work, will be defined in the conditions.

Contract time

The contract will normally specify a time within which the project must be completed, and the contract conditions will identify:

- how the project time commences;
- how completion of the project is recognised;
- damages to be paid by the contractor if the project is completed late;
- opportunities for the project time to be extended for certain delays outside the Contractor's control.

Payment

It is common on major contracts for payments to be made at intervals as the work progresses, both for the work specified in the original documents and for any changes. The amounts to be paid will usually depend on a number of factors, and exactly how the payment is to be calculated will be covered by clauses in the conditions.

Claims and their resolution

The construction of civil engineering works involves considerable risk, and most contract conditions will contain clauses identifying known risk areas and specifying whose responsibility these are. For example, where the weather is especially bad during the contract and because of this the Contractor is unable to finish on time, most conditions will contain a clause giving the Contractor the right to additional time for completion. For any risk that is accepted by the Promoter, the Contractor must make a claim for additional expense or for any extension of the project time resulting from these risks. Because there is often a dispute over whether a claim is justified, the conditions may also contain a procedure for resolving the dispute. Arbitration and conciliation are recognised methods of resolving disputes which are often included.

What are the options?

The conditions of contract are one of the contract documents which define the agreement between Contractor and Promoter and should be selected or prepared by the designer. The words 'selected or prepared' indicate that there is a choice here. In fact, the choice for the designer is between the following:

1. to write a set of contract conditions specifically for each contract;

2. to use an existing 'standard' set of contract conditions;

3. to use an existing 'standard' set of contract conditions and modify them to suit the particular contract.

Because conditions of contract are complex and take time to assimilate and understand, it is strongly recommended that a standard set of conditions be used. Option 1, above, is therefore *not* recommended. However, it is quite normal for the designer to specify (say) that the ICE *Conditions of Contract*, 6th edition (ICE6), are to be used on the project, but also to produce an additional document to amend or replace some of the clauses in ICE6.

A number of standard sets of contract conditions exist for construction contracts. Some are suitable for civil engineering, some for building. Some are for UK construction, some for overseas. Some are for bill of quantities-type contracts, some for other types. Some are for minor works, some for major works. Table 7.1 gives details of some of the most frequently used standard conditions in construction.

ICE *Conditions of Contract*, 6th edition (ICE6)

To be effective on construction sites, the people in charge, who will usually be civil engineers, must have a good understanding of the rules that apply; that is, of the particular conditions for their contract. Indeed, one of the reasons that experience is so valued in this area is the fact that familiarity with these rules that govern construction only comes with time. This means that the understanding that can sensibly be obtained through these notes is limited. The aim will be to make you aware of the kind of information that is covered by conditions of contract, so that, hopefully, you will know when to consult them.

Because ICE6 is still the most frequently used set of conditions in civil engineering in the UK, we will now concentrate on learning more about this particular document. As has already been said, ICE6 is not an easy read. It is written to be unambiguous, so that the words should convey a single meaning, and also it attempts to deal with complex issues. Any explanation will therefore be a simplification, and of course engineers on site dealing with contractual problems must consult the conditions that apply

Table 7.1 Frequently used conditions of contract in the construction industry

Full title	Shortened title	Area of construction	Type of contract	Comment
Conditions of Contract and Forms of Tender, Agreement and Bond for Use in Works of Civil Engineering Construction, 6th edition	ICE6	UK civil engineering	Bill of quantities	Probably the best known and most often used form in civil engineering
Institution of Civil Engineers Design and Construct Conditions of Contract		UK civil engineering	Design and build	Very new: developed because of the recent interest in design and build
Institution of Civil Engineers Conditions of Contract for Minor Works		UK civil engineering	Various	Less complex for simpler, less expensive projects
The New Engineering Contract	NEC	All areas	Various	A very new and innovative form which can be modified to suit a variety of projects
Conditions of Contract (International) for Works of Civil Engineering Construction with Forms of Tender and Agreement, 4th edition	FIDIC	International civil engineering	Bill of quantities	Available in several languages, although the English version is the official one
Joint Contracts Tribunal Standard Form of Building Contract	JCT	UK building	Bill of quantities	Several different versions of JCT conditions are available
General Conditions of Contract for Building and Civil Engineering – Standard Form of Contract	GC/Works/1	UK government contracts	Lump sum	May contain a BOQ, but the works are not remeasured

to the specific contract and interpret them accordingly: they must not attempt to apply these notes, other than as a general introduction. It should be noted that the terms Engineer, Contractor and Employer (using initial capitals) are used to mean the particular organisations fulfilling the roles defined in ICE6.

ICE6 contains 70 clauses, and the most important of these have been collected together loosely under general headings to try to distil a simple structure. The headings are not definitive, and certain clauses might be equally at home under more than one heading.

The parties on site

Figure 7.1 is a simple diagrammatic view of the main parties to be found on a typical ICE6 contract. The arrows show the principal lines of formal communication, and the double line clearly delineates the distinction between 'site' and 'headquarter offices'.

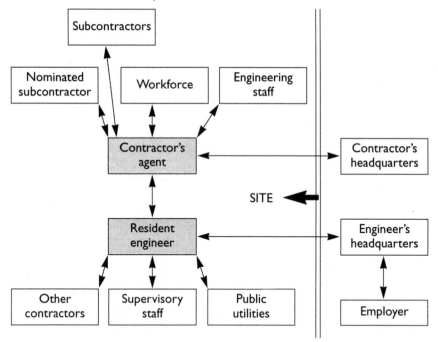

Fig. 7.1 Parties on site

Although the contract for construction is between the 'Contractor' and the 'Employer' (whom we have known so far as the Promoter), most of the clauses in ICE6 refer to the Contractor and the Engineer. The Contractor is the organisation that won the contract to construct the project. The Engineer must be a named chartered engineer working for the organisation that has been given the responsibility for supervising the project. It is normally expected that both the Contractor and the Engineer will have headquarter offices from which they will exert control over the contract.

The main figures on site are the Contractor's Agent and the Resident Engineer (RE), also known as the Engineer's Representative. On major construction sites, these two individuals will lead their own separate organisations to allow them to carry out their work, which, put simply, is for the Contractor's Agent to construct the Works and for the RE to ensure that all work is carried out properly. They will also be responsible for the work of other organisations involved in the construction process.

The Contractor's employees on site are the Agent, the Engineering staff and, of course, the workforce. The Agent and his/her staff must be competent (Clause 15) and have adequate knowledge to enable them to superintend the construction of the Works in a proper manner. The Agent is required to be 'constantly on the Works' and must be available to receive instructions from the Engineer or the RE. Some of the contract work will be carried out by the Contractor's own workforce, but other parts will be done by subcontractors. Clause 4 makes it clear that the Contractor is responsible for the work of the subcontractors just as much as for the work of his/her own workforce. Most subcontractors will have been selected by the Contractor, but occasionally the Contractor will be forced to work with a particular subcontractor who has been selected by the Engineer. Such subcontractors are known as nominated subcontractors and can cause considerable complications, as will be seen by the fact that two long clauses, Clauses 58 and 59, are devoted to them. Nevertheless, even when a subcontractor has been nominated, the Contractor will be responsible for the work of that subcontractor.

The Engineer's employees are the Resident Engineer and the supervisory staff, who will usually comprise civil engineers, possibly laboratory technicians and also inspectors, who are sometimes called clerks of works. The main function of this team is said, in Clause 2, to be to 'watch and supervise the construction and completion of the Works'. This means that the team will oversee the Contractor's work and will point out any instances where the workmanship or materials used are below the standard set out in the contract documents. In general, the inspectors will be in close contact with the work as it progresses, and will check line and level from the data established by the Contractor's engineers and checked by the RE's engineers. As an example of this co-operation, the way in which the laying of drains is typically inspected will now be explained. (Note that this is simply an example of good practice; ICE6 does not say how to supervise the laying of drains!)

Supervision of drainage

Drainage covers a multitude of activities, but here it is simply the laying of drains that will be addressed. Clearly the materials used must be checked for conformance with the specification, and this will cover the pipes themselves, the bedding material, pipe jointing systems and backfill material. It is the inspector who will check that these are either from an approved supplier or an approved source, and the engineers who will have arranged that approval. The materials must all be installed correctly, and this will involve checks that the proper depths of bedding have been used, that the pipes are in contact with the bed throughout their length and that the backfill is properly compacted. Once again, this is the inspector's job. It is, of course, all-important that the pipes be laid in the proper horizontal line at the design depths and to the design grades if the drainage system is to work efficiently. This will require the horizontal line of each drain to be set out by the Contractor's engineers and checked by the RE's engineers, and will also require that a system of level control be set up to ensure correct level and grade. Profiles are normally used for this purpose, with a profile board established at each end of the drain run in question. Figure 7.2 shows how such profiles are used.

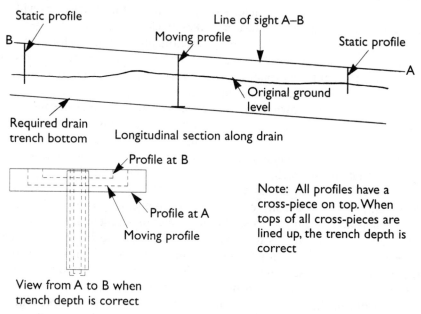

Fig. 7.2 Level control for drain construction

The profiles at each end of the drain run are established such that a line between the top of each cross-piece runs parallel to and a sensible distance above the line of the drain trench bottom. The moving profile is constructed so that the distance from the top of its cross-piece to its base equals the vertical distance between the line of sight from A to B and the drain trench bottom. Thus when the drain trench has been excavated to the correct level, the tops of all cross-pieces will line up. With a slightly shorter moving profile, it is easy to see how the line of the drain invert can also be established at all positions along the trench.

In general, it will be the engineers working for both Contractor and RE who will establish horizontal positioning and will set up the level control as just described. The workforce (for the Contractor) and the inspectors (for the RE) will be the people using these systems to check compliance.

It should be clear from all that has been said that the RE is given only negative powers to condemn bad workmanship or materials and has no authority to relieve the Contractor of any of his/her obligations or to change any aspect of the contract. The Engineer, however, has much greater powers, as will be seen in a later section, and Clause 2 states that some of these powers may be delegated to the RE. Where this happens, the Engineer must inform the Contractor of the powers that have been delegated. Certain of the Engineer's powers can never be delegated to the RE.

From Fig. 7.1, we see that there may be other contractors on the site, and often public utilities (water, gas, BT, electricity, etc.) will have their own contractors on site, diverting services out of the way of the main Works and installing new services where necessary. These contractors are not subcontractors, but will usually be working directly for the Employer and will be supervised by the Engineer. The main Contractor is not responsible for them. Under Clause 31, however, the main Contractor is expected to provide reasonable facilities for these contractors.

These, then, are the main parties on a typical construction site. The main players are the Agent and the RE, who together should be planning ahead to ensure that forthcoming activities are well organised and that the work is done properly at the first attempt. Inevitably, there will be discussions between subcontractors and the RE and between the public utilities' contractors and the main Contractor, but, as has already been said, the main formal lines of communication are as shown in Fig. 7.1. It would be most unwise to make agreements or give instructions that involve extra cost or delay along any other than these officially recognised channels.

The Employer's obligations

As has already been said, most of the clauses in ICE6 refer to the Engineer or the Contractor. The Employer, although one of the parties to the contract, gets little mention. It is clear, however, that the Employer does have certain obligations. These are:

1. to employ the Engineer, who will administer the contract.

2. to provide the site. Either this may be done all at once, or, if the contract has so specified, parts of the site may be made available on certain dates. Clause 42 makes it clear that if parts of the site are not available when they should be, any delays or extra costs the Contractor incurs may lead to claims against the Employer.

3. to pay the Contractor for completed work. Every month the Contractor

makes a claim for payment which the Engineer must check and change where necessary, and the Employer must then pay the Contractor the amounts the Engineer has certified (Clause 60).

4. to make available to the Contractor, prior to tender, all ground investigation information (Clause 11). The Contractor is then responsible for interpreting these data for the purposes of constructing the Works.

The Contractor's obligations

The Contractor's obligations are much more involved and include measures that attempt to ensure that the Contractor will produce a finished job, within time and also to the necessary standard. In detail, they are as follows:

1. The Contractor must *construct and complete the Works*, and must provide all the labour, materials, plant and temporary works necessary (Clause 8). It is also clear from this clause that the responsibility for *safety* of all site operations rests with the Contractor.

2. It is generally recognised that a visit to the site of the Works before any work is started can reveal much about the problems and difficulties that will be faced when work begins – problems that cannot so readily be detected from the drawings alone. Mainly for this reason, Clause 11 confirms that the Contractor is *assumed* to have made such a visit and *inspected the site* before preparing his/her tender.

3. To show how s/he intends to carry out the Works, the Contractor must *submit a programme* for the Engineer's approval (Clause 14). This programme must show the order in which the Works will be carried out but ICE6 does not say what format should be adopted, viz. CPM, time/distance chart, bar chart. Most contractors submit bar charts. If actual progress does not conform with this programme, a revised programme may be required by the Engineer.

4. Also under Clause 14, the Contractor may be required to submit *details of any temporary Works* to be used in the construction of the permanent Works. Figure 7.3 is an example of the temporary Works for the construction of a bridge. Although the organisation that completed the design (which may be the same organisation the Engineer works for) clearly designs the permanent Works, there may need to be a temporary support, often in the form of scaffolding, to allow the permanent Works to be constructed. These temporary Works, which can be complex and extensive, are usually designed by the Contractor. Under this clause, the Engineer is given the opportunity to inspect the Contractor's design to ensure that it is safe and that it will be adequate to produce satisfactory permanent Works.

Fig. 7.3 Temporary works

5. The position and orientation of the various parts of the Works should be able to be calculated and 'set out' from ground markers and bench-marks installed around the site. An example of setting out is shown in Fig. 7.4. Here, a base is to be constructed, and the centre of the base (X) will be established by intersecting lines of sight from adjacent ground markers using survey equipment. By then setting up over this point and taking a bearing from one of the ground markers, temporary markers (A, B, C and D) can be established outside the area of exca-vation for the base from which the main lines of the base can be deter-mined. When the excavation is complete, these temporary markers will be used to fix the formwork for the base and levels will be transferred onto the base from an adjacent benchmark. Normal procedure on most construction sites is for the Contractor's engineers to establish these temporary markers and for the RE's engineers to check them. Despite the fact that a check has been carried out, Clause 17 makes it clear that the responsibility for setting out rests with the Contractor.

6. There are a number of clauses (Clauses 19–25) that relate to the *safety of people and property,* and in general, the Contractor is made respon-sible. In particular, s/he is responsible for:

- the safety of all persons entitled to be on the site;

- the care of the Works;

- damage to persons and property resulting from the construction process;

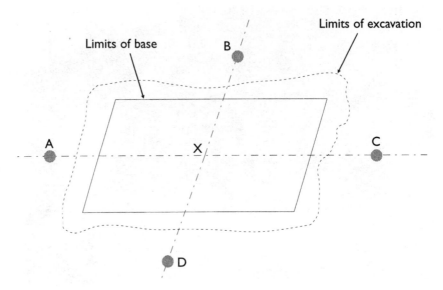

Fig. 7.4 Setting out of a base

and must *take out insurance* to cover any claims arising from these areas.

7. On completion of the Works, the Contractor must *clear the site*, leaving it in a workmanlike condition (Clause 33).

8. All *workmanship and materials* must conform with the standards laid down in the contract, and samples for testing must be made available as requested, before the materials are incorporated into the Works (Clause 36). The Contractor must also allow the Engineer access to all parts of the site and, in particular, must not cover up any work without the consent of the Engineer.

The Engineer's duties and powers

The Engineer's involvement in the contract is very extensive and will be reported under three headings: 'General', 'Claims and settlement of disputes' and 'Certificates'. Unless specifically mentioned, the powers described may be delegated to the RE.

General

The following are the Engineer's general duties and powers:

1. Where the Contractor has carried out work that is not up to the standards laid down in the contract, *the Engineer has the power to instruct the Contractor to remove the work* and/or require that it be properly redone (Clause 39).

2. A number of clauses deal with the time in which the contract must be completed, and Fig. 7.5 illustrates what is said. A time for completion of the works will have been specified, and one of the Engineer's first duties is to write to the Contractor and confirm the *date for commencement* (Clause 41). The Contractor is then expected to start work as soon as is reasonably possible after that date. If there are no delays to the contract for which the Employer is responsible, then the Works should be substantially complete when the time for completion has expired (*see* Fig 7.5a). If the Engineer judges that the Contractor is late in completing the works, the Contractor must pay damages to the Employer for each day or week the contract is late (Clause 47). These damages are known as *liquidated damages* and the level of payments will have been stated in the tender documents.

   Where there have been delays that the Contractor could not have envisaged, it would clearly be unreasonable to deduct damages without making some allowance for these delays. Clause 44 entitles the Contractor to an *extension of time* for completion of the works where:

- additional work is instructed by the Engineer (*see later*);

- there are increased quantities above those shown in the bill of quantities (*see later*);

- there is exceptional adverse weather;

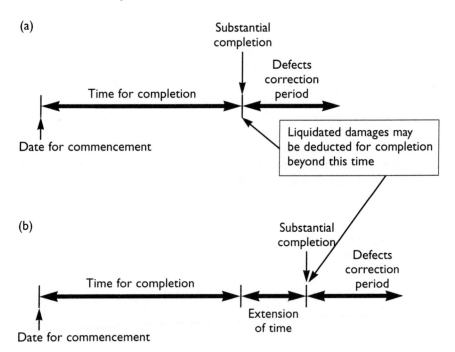

Fig. 7.5 Time in ICE6. (a) Contract time in the absence of delays; (b) contract time taking account of delays

- any other cause of delay occurs that is specified in the Conditions of Contract and that fairly entitle him/her to an extension.

In these circumstances, the Contractor must make a claim for an extension of time and the Engineer will assess whether an extension is justified and, if so, how long it should be. Where an extension is granted – and it may be requested and granted part-way through the job – this delays the point at which liquidated damages may be deducted (*see* Fig. 7.5b). *This power may not be delegated.*

3. The *progress* of all or any part of the Works *may be suspended* on the Engineer's written order and the Contractor must then protect and secure the suspended section. Depending on whether this suspension was included in the original contract, the Contractor may have claims for additional payments and possible extension of time (Clause 40). An example of an unplanned suspension would be the discovery of an uncharted gas main in a part of the site where construction is to take place. Here a suspension order would be needed to allow time for the public utilities to design and carry out a diversion of the service, and additional costs and delays would have to be considered.

4. If the Engineer at any time believes that the Contractor is *not making sufficient progress* to complete the Works on time, s/he may write to the Contractor to ask that steps be taken to ensure timely completion (Clause 46).

5. On a complex project, there will be many reasons why the original details in the contract documents cannot be constructed, and the Engineer has the power under Clause 51 to *vary the Works as necessary or desirable*. A simple example of such a variation is shown in Fig. 7.6 and stems from a mistake at the design stage.

   The figure shows a length of sewer to be constructed between manholes at A and B, but it is clear that with the design invert levels at the manholes and the actual ground profile between the manholes, the drain would be above the ground between a and b. Clearly this could not be constructed, nor would anyone want to construct it, and as soon as the error was detected it would be realised that the design would

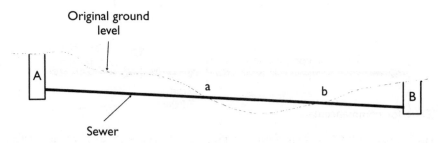

Fig. 7.6 Error in drainage layout

need to be changed and the Contractor instructed. This instruction may be given orally in the first instance, but should be followed up by a written site instruction. An example of a site instruction to correct the above defect is given in Fig. 7.7.

These changes or *variations*, as they are known in ICE6, will often involve increased costs, and *Clause 52 describes how variations are to be valued.* Where work is of a similar nature to work covered by the original bill of quantities (BOQ), then the rates in the BOQ are to be used (as stated in Fig. 7.7). However, where no similar rates exist, payment is to be made either by *daywork* or by *building up a rate* in line with the general level of rates and prices in the BOQ. (Note that daywork involves keeping a record of the plant, labour and materials used on the varied work and pricing these up at the daywork rates agreed.)

Two important points should also be made at this juncture:

- No variation is required where increases or decreases in the Contractor's workload are simply a result of the quantity of work actually carried out being greater than or less than that shown in the

---

Confirmation of Verbal Instruction

Site: Whickham Bypass

Section: Drainage

I confirm the following Site Instruction:

1.   Manhole A on drawing no. WB/R504/3 is to be constructed as a type 2/300/B with a revised invert level of 35.425 m.

2.   Manhole B on drawing no. WB/R504/3 is to be constructed as a type 2/300/B with a revised invert level of 33.216 m.

3.   The connecting drain between manholes A and B is to be constructed as a 300 mm carrier drain type D, with revised invert levels of 35.425 m at A and 33.216 m at B.

Method of Payment: BOQ Part 3.1.10.5 Items 22 & 27

Resident Engineer

Fig. 7.7 Site instruction

original BOQ – that is, where no change to the Works has been made, but the actual quantities are different from those shown.

- A variation order is a document stating the way in which work has been altered and the revised payments to be made as a result. Because the costs of variations are often not agreed before the work is done, the variation order is usually prepared at a later date.

6. Clause 55 makes it clear that *the quantities in the BOQ are estimated quantities only*, and we will see later that the Contractor is paid on a monthly basis for work done up to the end of each month. It is the Engineer's job to determine how much the Contractor should be paid by *admeasurement* (Clause 56); that is, by remeasuring the actual work done. The Contractor is requested to attend when the work is remeasured, hopefully leading to agreement of quantities. This remeasurement is aimed at paying the Contractor for work done; however, it should be realised that if a wall is specified to be built 2.5 metres high, say, and the Contractor builds it 2.7 metres high by mistake, the Engineer might accept it, but would only pay for the height of wall specified.

7. It has been explained above that ICE6 recognises that the quantities in the BOQ are not always correct, and Clause 56 allows the Engineer to *revise a rate where there is an increase or decrease in the actual work done compared to the estimates* that renders the quoted rates unreasonable. For example, the BOQ might state that there is 10 000 m^3 of excavation, but it is later realised on remeasurement that there is only 2 000 m^3 of material to be excavated. In such a case, the Contractor might well argue that the rate s/he placed in the BOQ was low because of the large quantity, and that with a much reduced quantity, the rate would have been much higher. Here, Clause 56 would allow the Engineer to consider the Contractor's arguments for an increased rate and make an adjustment if the case was seen to be reasonable.

Claims and settlement of disputes

There are several clauses in the *Conditions of Contract* that give the Contractor a right to claim additional costs and, in some circumstances, additional time. We have already been introduced to Clause 44, which covers claims for additional time, but it is Clause 52(4) that explains *how the Contractor must proceed to make a claim for additional payment*. In general, the procedure is as follows:

1. The Contractor must give notice of the intention to claim as soon as s/he becomes aware of the circumstances that are believed to justify a claim.

2. The Contractor must then keep contemporary records to support the claim.

3. As soon as possible, the Contractor must submit an account detailing the grounds for the claim and the amounts being claimed. This account may be updated at intervals as the full extent of the claim becomes clear.

4. If the Contractor complies with the above procedure, s/he is entitled to payment on the claim to the extent that the Engineer considers is due.

Where there is a dispute between the Employer and the Contractor arising out of the contract, it is Clause 66 on the *settlement of disputes* that must be followed. Such disputes might result from disagreement about the Engineer's decisions on rates for varied work or on amounts the Engineer considers should be paid against other Contractors' claims. Again a procedure must be followed:

1. Most disputes will be originated by the Contractor who is not satisfied with the level of payments the Engineer is making, and the first step s/he must take is to serve on the Engineer a *Notice of Dispute* stating the nature of the dispute.

2. On receipt of this notice, it is the Engineer who makes a decision about the dispute and informs the Contractor and the Employer in writing. *This power may not be delegated.*

3. If either party is still not satisfied, then they may take the dispute further; either to conciliation or to arbitration. Put very simply, *conciliation* involves a mutually agreed third party listening to the cases made by both sides of the dispute and making a recommendation which will hopefully be acceptable. *Arbitration* is usually a much more expensive method of resolving disputes and is similar to a court of law, except that the arbitrator will be an expert in construction and that the arbitrator's final decision is not made public. The arbitrator's decision in most cases is final and binding on both parties.

As an example of a contractual claim, consider a site on which there is a large excavation in material that the boreholes show to be relatively soft. During the work, a sizeable volume of harder material is found but there is no item in the BOQ for excavation in rock. The Contractor claims under Clause 12 that the rock constitutes a '*physical condition that could not reasonably be foreseen by an experienced contractor,*' and begins to keep records of what s/he says are additional costs. These might involve the use of pneumatic tools or explosives, or a reduced productivity with existing plant. The Engineer, who will eventually receive claims for additional payment from the Contractor, must first decide whether a claim is justified. There may be some doubt, as the actual hardness of the harder material may not be so much harder than the material that showed up in the boreholes. If the Engineer does accept the basis for claim, s/he may not accept that the amount of money the Contractor is claiming can be justified. A decision must, however, be taken and payments made to the Contractor.

If the Contractor does not accept the Engineer's judgement on the monies due, s/he must decide whether to serve the Engineer with a Notice of Dispute, which may lead to conciliation or arbitration.

Certificates

There are a number of certificates that will be issued at particular points in the progress of the contract and Fig. 7.8 shows these diagrammatically.

At the end of each month, the Contractor submits to the Engineer a statement detailing the money s/he considers due (often called a valuation). The Engineer must adjust the valuation as s/he sees fit and certify the actual amount payable in an *interim certificate* (Clause 60). This payment must be made to the Contractor by the Employer within 28 days of receiving the valuation. An interim certificate will usually include:

- payment for original contract work measured by BOQ items;

- payment for varied work using BOQ items;

- payment for varied work on a daywork basis;

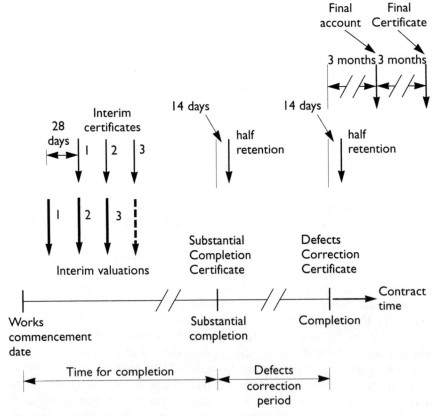

Fig. 7.8 Certificates required by ICE6

- payment for varied work using 'built-up' rates (interim rates may be used while negotiations are still ongoing on the final rate to be adopted and these are sometimes known as * rates);

- a percentage may be payable for 'materials on site' which are not yet incorporated into the Works;

- interim payments on claims may be made;

- where a 'contract price fluctuations' clause is included in the contract, a figure for the increase or decrease as a result of variation in construction indices will be calculated each month and added or subtracted. (This is a mechanism that allows the Contractor to give prices in the tender that relate to the time of tender and for these rates to be adjusted automatically each month to allow for inflation.);

- where an 'adjustment item' has been used (*see* Chapter 4), a percentage of the adjustment item amount will be added or deducted in proportion to the amount of work done.

All of these sums represent the total work done up to the date of the valuation, and to determine how much the Contractor should be paid this month we must deduct the amount paid last month. Another deduction is also made, known as 'retention'. This is usually 3% of the sum due to the Contractor and is held back in case the Contractor fails properly to fulfil the contract. It is evident from Fig. 7.8 that this money is returned to the Contractor in two parts at different stages.

When the Contractor considers that the whole of the Works are substantially complete, s/he can give notice to that effect to the Engineer and agree to complete any outstanding work during the *defects correction period*. This is then seen as a request for the Engineer to issue a *Certificate of Substantial Completion*, which s/he will do if the Works are thought to be substantially complete. *This power may not be delegated.* The normal procedure is for the RE to tour the site to make a 'snagging list' of all outstanding work. If there is nothing major on this list and if the Employer can sensibly make use of the constructed facility, the Engineer will issue the certificate.

During the Defects Correction Period that follows the issue of the Certificate of Substantial Completion, the Contractor must:

- complete any outstanding contract work;

- remedy any defects resulting from unsatisfactory workmanship or materials that become evident during the period;

- carry out repairs as instructed by the Engineer, for which additional payment may be made.

At the end of the Defects Correction Period, provided all repairs have been made good, the Engineer will issue a *Defects Correction Certificate*.

This certificate signals that the Contractor's main obligations to construct the Works are now completed. *This power may not be delegated.*

Within three months of the date of the Defects Correction Certificate, the Contractor must submit to the Engineer a *statement of final account* detailing all the sums s/he considers to be due under the contract. The Engineer then has a further three months in which to issue the *Final Certificate* showing the amounts that s/he considers are finally due for the work carried out. The balance must then be paid within 28 days. *This power may not be delegated.* The Final Certificate will include similar items to the interim certificate, with the following exceptions:

1. All of the BOQ work will have been completely remeasured and final quantities agreed.

2. Any * rates for varied work will either have been agreed or 'fixed' by the Engineer.

3. There should be no 'materials on site' element.

4. Claims payments, contract price fluctuation and adjustment item payments should all be final figures.

5. No retention will be deducted.

Final note

The aim of this section is to give you some awareness of the rules that govern most construction contracts during the construction phase and for you to understand how those rules affect the way that engineers must behave. If in the future you are involved in contracts not governed by ICE6, you will need to study the particular conditions of contract that apply. That does not mean that what you have learnt here is wasted, as most conditions will need to address similar issues and will often do so in similar ways. Do, however, remember that the understanding gained from this short section is only an overview and that there is much more to learn.

Recommendations for further reading

Institution of Civil Engineers (1991) *Conditions of Contract,* 6th edition (ICE6). Thomas Telford, London.
Abrahamson, M.W. (1979) *Engineering Law and the ICE Contracts*, 4th edition. Applied Science Publishers, London.

To understand the kind of language used in conditions of contract, the document itself (ICE6) should be investigated, and to gain some understanding of how specific clauses have been interpreted, Abrahamson is most instructive. It should be noted, however, that Abrahamson's book actually addresses the 5th edition of the ICE *Conditions of Contract*, and for some clauses in the 6th edition will not be relevant.

8 Safety

Introduction

Construction sites are dangerous. Nobody in the industry doubts this statement and there are serious attempts being made to reduce the number of accidents, which occur far too frequently in these temporary and constantly changing workplaces. Professional engineers, working either for the contractor or for the supervisor, should be particularly aware of the kind of risks most likely to cause accidents and will certainly not want to feel that they have been in any way responsible for another's injury or death as a result either of their actions or, indeed, of their inaction. The responsibility for safety at work, however, is not just left to the individual's conscience and humanitarian feelings as there is statutory legislation that identifies the specific responsibilities of both employers and employees (Health and Safety at Work Act 1974). Individuals may be prosecuted under this legislation for failing to behave in a responsible manner and the prosecutions may result in fines or, in particularly severe cases, imprisonment.

One of the latest pieces of safety legislation to affect the construction industry, the Construction (Design and Management) Regulations 1994, recognises that opportunities for unsafe working may actually originate at the design stage. This legislation thus requires the designer to consider the safety implications of his/her design and to avoid or attempt to reduce any foreseeable risk to the health and safety of those involved in construction and also those who may later be required to maintain the constructed facility.

The young engineer thus needs to be well prepared before starting work, either in a design office or on a site. This preparation should involve a good understanding of the inherent dangers that exist on many of our construction sites, coupled with an awareness of the current safety legislation to ensure that s/he is operating within the law. Perhaps more important than all of these, however, is a proper attitude towards safety – an attitude that is constantly looking to identify risks to health and safety and attempting to remove or minimise those risks.

The extent of the problem in the construction industry

Accidents on construction sites vary considerably in their severity. Some only cause the injured party to be away from work for a few days, while others will result in serious (major) injury which may involve permanent disability – and there are, of course, accidents which are fatal. Alongside these incidents, whose effect is normally immediate, there are what are sometimes known as the 'slow accidents'. These may cause problems a long time after the actual incident that caused the injury occurred. An example of a 'slow accident' would be the inhalation of asbestos dust, which may lead to asbestosis, a scarring of the lungs causing shortness of breath and long-term health problems.

Details of accidents occurring in British workplaces are collected by the Health and Safety Executive (HSE) and categorised by different industries and by different accident type. One of the ways in which safety statistics are presented is in the form of number of injuries per 100 000 employees, and in the recent past these figures for fatal and non-fatal injuries have been gradually reducing. This trend is shown in Figs 8.1 and 8.2 (derived from HSE data), where the statistics for both the construction

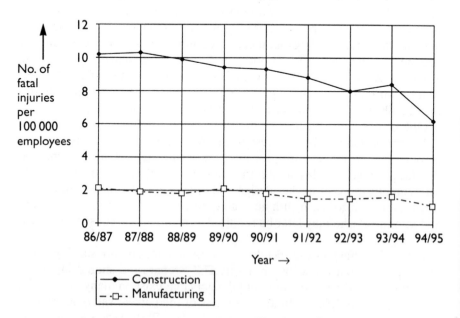

Fig. 8.1 Fatal injury rates per 100 000 employees: 1986/87 to 1994/95

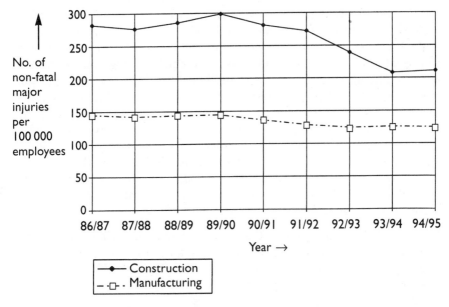

Fig. 8.2 Non-fatal major injuries per 100 000 employees: 1986/87 to 1994/95

industry and manufacturing industry appear as separate lines. These statistics should not, however, lead to any complacency. The fatality rate in construction is still several times the rate we find in the manufacturing industries and the major injury rate is consistently higher in construction. To give some figures for the construction industry rather than just quoting rates, in 1993/94 a total of 73 employees, 15 self-employed workers and 6 members of the public suffered fatal injuries, while some 1801 employees, 766 self-employed and 116 members of the public suffered major injuries. Each of these incidents will have caused varying levels of grief and trauma to the injured and his/her family, and, of course, the economic arguments of lost productivity, lost income and the drain on the health service should also be recognised.

The main causes of accidents

It is clear from the evidence in the previous section that the construction industry is more dangerous than manufacturing industry, and it is not too difficult to understand why that should be. Most manufacturing will take place in a factory which has been designed and laid out for the production of specific goods and the layout will have been planned with effective and safe production in mind. Construction is quite different. The workplace for construction workers will typically be in and around the facility being constructed, which must, by definition, be constantly changing. Actions to ensure safe access and safe working areas must therefore be regularly reconsidered as the construction proceeds, otherwise safety will be compromised. It is also true that much of the work of construction is carried out by small firms and self-employed workers who may not be well versed in safe working practices. All will be working to a price which will usually be competitive and unlikely to contain a high profit margin. In these circumstances, safety may be thought to be less important than the bottom line. Where the pressure to behave irresponsibly is great, it is clear that legislation must enforce sensible procedures, and this is what the safety legislation attempts to do. There are, however, many construction sites and relatively few safety inspectors to visit them. Day-to-day safety must therefore be controlled by those responsible for the work, who must be well aware of the principal sources of risk. From the experience of the HSE, the major accident areas in construction, in order of frequency of occurrence, are as follows:

1. *People falling.* This is easily the most frequently occurring type of accident and covers falls from scaffolds, ladders, down stairs, from roofs, into excavations and even falls on the flat. Typically, 50% of all construction fatalities are caused by people falling.

2. *Being struck by falling objects.* Objects falling onto people are the next most frequent category contributing to fatal injuries. The objects may fall from scaffolding, during lifting operations, off vehicles or as a result of demolition or collapses. Wind-blown materials are also known to cause injury.

3. *Vehicles/plant/machinery.* Vehicles used for transporting people and materials, together with stationary mechanical plant, cause several fatal injuries each year. These may involve vehicles reversing, overturning or colliding and running over or trapping people against obstructions, while stationary plant may involve accidents where people are trapped by moving parts.

4. *Excavation/tunnelling.* Construction of foundations and drain trenches will involve temporary excavations which if not properly supported may collapse and trap or bury workmen. Excavated material placed at the edge of the excavation can effectively increase the depth of the hole. Collapse during tunnelling is likely to be even more catastrophic and it is essential to keep the unsupported area next to the face to a minimum.

5. *Electricity.* The effect of electrical shock on the human body can be fatal if the current is high and/or suffered for an extended period, but it may also be very serious if the shock leads to a fall. Exposure to electrical current may stem from contact with overhead or buried cables or may simply occur by touching faulty electrical appliances.

6. *Fire/explosion.* Many site operations involve the use of flammable materials, including liquid petroleum gas, and without proper control the

scope for fire and explosions is evident. The most common accidental fires on construction sites, however, occur not in the facility being constructed, but in the temporary site huts and stores.

7. *Asphyxia/drowning.* The problems of asphyxia are, of course, most likely to occur in situations where work is being carried out in confined spaces and fresh air is restricted. Unfortunately, much construction work falls into this category. Trenches, tunnels, boreholes, box girders and manholes are just a few examples of areas where construction must take place and where there is a risk of inadequate ventilation. Accidents have occurred where workers trying to help victims of asphyxia have entered the confined spaces themselves and suffered the same fate. Any work in or near water, where the worker is not paying sufficient attention to the imminent danger, clearly involves a risk of drowning.

Safety legislation

The Construction Regulations

Although safety legislation existed previously for the building industry, it was not until 1962 that the safety of workers in the civil engineering side of the construction industry was covered by legislation. By 1966, there were four sets of relevant regulations made through powers conferred by the Factories Act. They were the Construction (General Provisions) Regulations 1961, the Construction (Lifting Operations) Regulations 1961, the Construction (Working Places) Regulations 1966 and the Construction (Health and Welfare) Regulations 1966. These remained in force for many years, but three of them have now been superseded by the Construction (Health, Safety and Welfare) Regulations 1996 (CHSW) and the fourth, on Lifting Operations, is due to be superseded in 1997. These changes have been made and continue to be made to comply with directives on safety from the European Union. The new regulations that have been brought into force in this way will be dealt with at the end of the chapter.

The old form of the Lifting Operations regulations, still in use at the time of writing, covers the use of lifting appliances and lifting gear on construction sites. Such plant and equipment is to be kept in good order and safe for use, and is required to be used safely. The regulations also address such issues as inspection and testing and require that safe working loads are marked on lifting gear.

Health and Safety at Work Act 1974

In 1974, the introduction of the Health and Safety at Work Act (HSW Act) marked a change in the approach adopted for implementing safety legislation. Previous legislation had concentrated mainly on the need for premises to be safe and paid particular attention to how that could be achieved, whereas the HSW Act puts the emphasis on individuals and their duties. The main aims of the Act, which covers all people at work, were to secure the health, safety and welfare of the workforce and also to protect members of the public from the effects of people at work. This was to be achieved by specifying the duties of employers, employees and also of those who designed and manufactured products for use at work.

Under the Act, the *employers* who clearly have the main responsibility, being in control of the organisation of construction, have a duty to ensure the health, safety and welfare of their employees and others and are required to discharge this in a number of ways. They must provide and maintain safe plant and systems of work, safe methods of handling dangerous substances, safe workplaces with adequate welfare facilities, and instruction and training where necessary. For all but the smallest of organisations, the employer must also provide a written statement of the health and safety policy of the organisation and make all employees aware of this.

Designers and manufacturers of products for use at work – for example, companies that produce scaffolding equipment – are also given responsibilities under the Act. They must ensure that their products are safe for the purposes for which they are intended, carry out any tests needed to prove this and provide adequate information about the correct use of the equipment.

Employees, by their own careless acts, can create circumstances in which either they or others may come to harm, and this is recognised in the Act. They are expected to take reasonable care to avoid such occurrences and must co-operate with their employer as necessary to enable the employer to comply with statutory duties.

As part of the HSW Act, the Health and Safety Commission was established as the prime authority responsible for all health and safety at work matters. Its main functions are to provide information and advice on safety and to develop proposals for the reform of the existing regulations. It therefore determines safety policy for industry, but also issues codes of practice to provide practical guidance on the requirements of the HSW Act.

The Health and Safety Executive operates under the direction of the Commission and its main job is to enforce the requirements of the Act. In construction, this is done through inspectors working for the Factory Inspectorate, who have the legal right to make unannounced visits to construction sites to check whether safe practices are being observed. Where problems are identified, the inspector has a choice of three main courses of action and will decide how to proceed depending on the nature of the particular problem. If the contravention appears to be an oversight and is not likely to cause imminent danger, the inspector may give verbal advice to allow the fault to be corrected. If the inspector considers that there has been contravention of a statutory provision which is likely to continue or to be repeated, s/he may issue an *improvement notice* which requires the contractor to remedy the situation. Where the inspector believes that there is a risk of serious personal injury, a *prohibition notice* may be issued requiring that the activity in question be stopped until the contravention is remedied. The inspector's final recourse, when confronted by a particularly intransigent individual or organisation, is to prosecute. If the defendant is then found guilty, a fine or a prison sentence or both may result.

The HSW Act is the primary safety legislation in the UK and it is through powers conferred by this Act that a considerable amount of new safety legislation has been made. As mentioned previously, some of the new regulations have been brought into force to comply with European Union directives, and the ones that have a direct application to the construction industry will now be described.

Management of Health and Safety at Work Regulations 1992

These regulations, which apply to all businesses in all areas of work, require, amongst other things, that employers make a specific assessment of the risks to the health and safety of employees and others resulting from activities carried out at work. The assessments, which must be recorded for all but the smallest organisations, are intended to identify any preventive or protective measures that are necessary. Any such measures must then be implemented by competent people.

Provision and Use of Work Equipment Regulations 1992

In these regulations, all the existing laws relating to equipment used at work are combined. Employers and the self-employed are required to ensure that suitable equipment is provided and that it is properly maintained with information and training available, as needed.

Personal Protective Equipment at Work Regulations 1992

A duty is imposed on all employers to assess the need for protective equipment and clothing and to provide any that is required. In the construction industry, where workers are often outdoors, clothing to protect from adverse weather will frequently be needed. Safety helmets will also need to be provided. The employer has the additional responsibility of trying to ensure that the equipment provided is actually used.

Manual Handling Operations Regulations 1992

Hazards and risks associated with the movement of loads by hand must be assessed by employers and any action necessary to avoid or reduce the risk of injury must be taken. This regulation is particularly relevant to the construction industry, which relies heavily on this kind of activity.

Construction (Design and Management) Regulations 1994

Having designed a civil engineering project and effectively defined the work to be carried out, the safety issues must then revolve around how that work is to be executed in a safe manner – the main responsibility of the Contractor. If this is all that is done, however, opportunities for affecting the safety of the project by considering safety at the design stage will have been overlooked. The Construction (Design and Management) Regulations (CONDAM) recognise that there are considerable advantages to be gained from taking safety into account at the design stage and identify an obligatory procedure for doing this. All projects which include construction work, other than very minor works, must now adopt this procedure, which places duties on the *client*, the *planning supervisor*, the *designer*, the *principal contractor* and *other contractors*. It also requires the production of a *health and safety plan* and a *health and safety file* for every project. Because the health and safety plan and the health and safety file figure so prominently in the duties of the nominated persons, they will be described first.

The health and safety plan

The health and safety plan is prepared initially by the *planning supervisor* and should be part of any tender documentation so that prospective principal contractors are aware of its contents when preparing their tenders. It is intended to identify the principal risks in the project and an indication of the precautions for dealing with them. At this point, the plan should contain, amongst other things:

- a description of the work in the project and the time-scale within which the work is to be carried out;

- information about the existing environment, in and around the site;

- details of risks to the health and safety of those who will construct the work to the extent that they can be reasonably foreseen together with recommended precautionary measures;

- principles of any structural design and any sequences of assembly that need to be followed.

The principal contractor then takes over the plan and must develop it until the end of the construction phase by making it specific to the approach that s/he intends should be adopted by everyone involved in the construction phase. This should include:

- assessments of risk carried out under the Management of Health and Safety at Work Regulations;

- common arrangements for all contractors on the site, typically for emergency procedures and welfare;

- arrangements to ensure that all contractors and their employees comply with the rules contained in the health and safety plan.

The health and safety file

The health and safety file is intended to provide information about the project to those who may do work on it at a later stage, which might include its repair, maintenance or demolition. The planning supervisor must ensure that the file is prepared and handed over to the client on completion, and is likely to need the assistance of the principal contractor in this. It is likely to include:

- as-built drawings, construction methods and materials;

- details of maintenance facilities and procedures, including any specially prepared manuals;

- details of services and any emergency and/or fire-fighting systems.

The parties involved

The *client*, the person for whom the project is being carried out, must appoint a planning supervisor and a principal contractor, and for most civil engineering works will need to appoint a designer, all of whom will play major roles at different stages in the management of safety for the project. These nominated people must be competent and able to allocate sufficient resources to the project and it is the client's responsibility to ensure this.

The role of *planning supervisor* is a completely new one for the industry and s/he will play an important part in the project from the design stage through to the completion of construction. We have already heard how s/he is involved in the production of the safety plan and the safety file, but the planning supervisor must also monitor the health and safety aspects of the design and be available to give advice to the client or to any contractor.

The CONDAM regulations place a responsibility on the *designer* to consider the risks to those who will construct the project and those who will maintain it. Where possible, risks should be avoided, but if this is not possible, the designer should endeavour to reduce the risks. Any information about the design which might affect health and safety should also be provided.

The part that the *principal contractor* must play in producing the safety plan and the safety file has already been described, but there is more that s/he must do, mainly in a co-ordinating role. It is recognised that there may be several contractors on the site, and the principal contractor must take reasonable steps to ensure that they work together effectively to comply with the safety legislation and to create a safe environment.

Construction (Health, Safety and Welfare) Regulations 1996

The Construction (Health, Safety and Welfare) Regulations (CHSW) only came into force in September 1996. As has been mentioned, they replace three of the sets of regulations made in the 1960s, and contain several regulations that cover the health and safety of all those involved in construction work by the use of general statements indicating safe practice. For example, Regulation 5 states:

There must be safe access to and from the place of work. All workplaces must, as far as reasonably practicable, be kept safe and without risks to health. Workplaces should be arranged to allow sufficient working space and to be suitable for anyone who is working or is likely to work there.

Similar defining statements are also made about the following:

- prevention of falls
- protection from falling objects
- safety of existing structures adjacent to construction work
- safe demolition
- use of explosives
- collapse of excavations
- cofferdams and caissons
- avoidance of drowning
- moving vehicles
- fire, explosion and asphyxiation
- working temperature and provision of protective clothing
- adequate lighting
- safe, well maintained plant and equipment
- need for trained personnel
- inspection of excavations, cofferdams, etc. by competent persons
- traffic routes on construction sites
- emergency arrangements on construction sites

- fire-fighting equipment on construction sites
- sanitary and washing facilities on construction sites.

The duty to comply with these statements rests with employers, the self-employed and all who control the way in which construction work is carried out. Employees, however, are also required to comply with any of the regulations that apply to them.

Final note

It should be clear from all the above that there will be areas of overlap where more than one set of regulations appears to apply to a particular situation. This is recognised, and where it happens it is said to be generally acceptable to comply with the regulations that apply most specifically. It should also be clear that the area of safety law is extremely complex. Only the briefest of outlines has been given here to act as an introduction, but the engineer with responsibilities on a construction site should receive much more detailed information from his/her employers and would be well advised to pay careful attention.

Recommendations for further reading

Health and Safety Commission (1995) *Management of Health and Safety at Work Regulations 1992 Approved Code of Practice.* HSE Books, Sudbury. 25 pp.
Health and Safety Commission (1996) *Managing Construction for Health and Safety: Construction (Design and Management) Regulations 1994 Approved Code of Practice.* HSE Books, Sudbury. 42 pp.
The codes of practice published by the Health and Safety Commission are much more readable than the legislation they address and give a detailed interpretation.

Davies, V.J. and Tomasin, K. (1996) *Construction Safety Handbook.* Thomas Telford, London. 320 pp.
A general book on safety, including legislation up to and including 1996.

9 Estimating and tendering

Introduction

In the modern world, it is difficult to imagine a construction project that is not cash-limited; that is, with no restrictions on the funds available to finance it. Public-sector projects must compete for funds with alternative uses of tax-payers' money and must therefore operate within tight budgets. The private sector is also most unlikely to be in the position of wanting a particular project regardless of the price. This means that we invariably want to know the final cost of our projects, not only before we begin construction, but before we commit money to detailed design. The need to forecast project costs at an early stage means that the business of estimating is a major concern in the construction industry, and although we would very much like to be able to produce good, accurate estimates, as we will see this is very difficult to achieve.

The next section will show how estimates are produced at different stages of a project's life, but for the moment it is important to make the distinction between estimates and tenders:

- An *estimate* for a civil engineering project is an attempt to predict the cost of the project at some stage before its completion.

- A *tender* is the price at which a contractor offers to complete the project and will be based on a careful estimate of the project cost.

Quite how a tender differs from an estimate will be explained in more detail in a later section.

Estimates at different stages

Table 9.1 shows a summary of the three main stages during a typical project when estimates will be made.

Table 9.1   Stages at which estimates are produced

Stage	Estimate produced by	Reason for estimate	Methods used
Feasibility study	Designer	To compare alternatives and give the promoter an indication of the funds required	Comparisons with previous similar projects or use of composite rates viz. rate/m^2 of road construction or office space
Completion of detailed design	Designer	To provide a more accurate indication of the funds needed to complete the project and to assist in the assessment of tenders	Usually, insertion of BOQ rates from previous similar projects, adjusted as necessary, into the completed BOQ for the current project
Tender stage	Contractor	As the basis of the tender	Operational estimating Unit rate estimating Subcontractor's quotations

The first estimate will be produced as part of the feasibility study and will usually be one of a number of estimates of each of the possible alternative projects being investigated. The relative cost of each scheme will be an important factor in deciding which project should be selected and will also give the promoter an understanding of the general level of funding needed to finance the project. With this information, the promoter may decide to proceed to detailed design and eventual construction, or may 'shelve' the scheme until sufficient funds are available. The accuracy with which any scheme can be estimated at this stage is low, since the designs will only specify an outline of what is being considered. The main aims should be to produce estimates that are capable of distinguishing between the alternative schemes studied and for the promoter not to get a nasty shock later in the project's life when the contractors' tenders are received.

Because there will be little detail to define each project, the methods of estimating costs must necessarily be imprecise. Established firms of consulting engineers may well have designed projects similar to the current one for previous clients, and the final costs of these, adjusted for inflation, size and any other important aspects, will give an indication of the cost of the present project. Alternatively, the consultant will usually have an updated databank of composite rates. For example, in estimating the cost of a warehouse or office, this may be done on the basis of a rate per square metre of warehouse space or office space that can be applied to the present project. The estimate may be adjusted for particularly difficult foundation conditions or any other factor that would make the present project unusually cheap or expensive.

The second estimate will also be produced by the firm that designs the project, but this estimate will be made when the detailed design is completed. With precise details of exactly what the project entails, the methods of estimating can be more accurate, and where a bill of quantities (BOQ) has been produced, the estimate will usually be carried out by entering rates into the tender BOQ and computing the total. The problem, of course, is knowing what rates to enter, and once again established consulting engineers will rely on data they have built up from previous contracts they have been involved with. For this estimate, the historic data used will be in the form of bill rates and an entry in the consultant's database might appear as shown in Table 9.2.

Table 9.2   Example of entry in a rate database

CESMM3 code	Item description	Average rate	Individual rates	Quantities	Contract
F624	Placing of reinforced concrete in bases, thickness exceeding 500 mm	10.88	10.30 16.20 7.45 9.55	20 11 348 165	C121 B095 C82 C163

If we have a similar item to price in the current project, this information shows us what prices have been used for this item in previous contracts. If it is assumed that inflation factors have been automatically applied to bring all of these rates up to present-day rates, we still have a problem in selecting a rate. Should we use the average rate? If not, what rate should we use? It is evident from the database that different rates were offered on different contracts which included different amounts of this item to be carried out. The information about the actual contract on which these items were priced may also affect our decision if we know that the concrete on certain contracts was particularly difficult to place. It should therefore be clear that producing the estimate at this stage is not simply a mechanical process. Judgement and knowledge are needed to make sensible selections and more care should definitely be taken on those items that are particularly expensive.

When we have produced this second estimate for the project, for what purposes will it be used? The obvious purpose is to provide a more accurate forecast of the project's final cost and allow the promoter to ensure either that sufficient funds will be available or that adequate budgets have been set. It can also be helpful in assessing contractor's tenders when they are received. If six contractors have been asked to tender for the job and their tenders are as shown in Table 9.3, the promoter will typically want to offer the contract to contractor C, who made the lowest offer.

Table 9.3 Contractors' tenders

Tenderer	Tender (£ million)
A	7.62
B	7.49
C	6.10
D	7.86
E	8.68
F	7.24

If, however, the consultant's estimate suggests that the minimum reasonable tender is £7 million, this would suggest either that contractor C has made a mistake in developing his/her tender or that s/he has misunderstood the requirements. It might be wiser to accept the tender of contractor F in such circumstances.

The third estimate will be produced by the contractor and will be the basis of his/her tender. If successful, the contractor must construct the works and, for a bill of quantities contract, be paid in accordance with the contract at the rates inserted in the bill. The estimate must therefore be sufficiently competitive to allow the contractor's tender to win the bidding competition while also ensuring that the contractor's costs are covered and that, hopefully, some profit is made. It is this estimate and tender that will now be addressed in detail in the rest of this section.

The contractor's tender

We have already heard how the tenders of competing contractors will be used to select one contractor to construct the works, and in the very first section of the book, the tender period, during which each contractor must study the tender documents and decide on his/her offer, was described. Now we need to understand more about exactly how the contractor puts the offer together and how these most important decisions are made. Use of the word 'important' here might even be an understatement. If a contractor's tenders are too low, contracts may be won on which money is actually lost; if tenders are too high, no contracts will be won. Neither of these scenarios could be sustained for long without the contractor suffering serious financial problems.

It has already been said that the contractor's tender needs to be based on a careful estimate of the project costs, and the relationship between the two is usually shown as:

Tender price =   Estimate   +   Mark-up

where:

Estimate   =   Direct costs   +   On-costs

Mark-up   =   Company   +   Risk   +   Profit
                 overheads

Thus the estimate consists of direct costs and on-costs.

Direct costs

Direct costs are the principal costs of carrying out the work of the contract, viz. excavation, foundations, structural work, etc., and will comprise the costs of employing labour, of providing materials and plant and of any temporary works necessary. Any wastage of materials and idle time should also be allowed for here. These costs are probably the most difficult to forecast properly and are the costs to which the methods of operational estimating and unit rate estimating will be applied.

On-costs

On-costs are sometimes known as site overheads and consist mainly of the cost of the contractor's site management team, their transport and their accommodation. All of these are essential to enable the contractor to construct the works and yet are not covered in the direct costs.

The mark-up consists of company overheads, risk and profits.

Company overheads

A contractor's main source of funds is the money made from the contracts undertaken. The cost of running the main headquarters organisation where administrative and estimating functions are performed must therefore be provided by contributions from each contract won. This is shown in Fig. 9.1, where the London office and three regional offices of a major contractor are being supported by money made on its sites. Thus a percentage is added to each estimate to cover these costs.

Risk

In agreeing to carry out the works as specified at a given price, the contractor is taking a risk on the basis of his/her judgement of many factors

Fig. 9.1 Contribution of sites to company overheads

and experience of similar work. Depending on how damaging the failure to judge properly all the relevant factors would be in financial terms, the contractor may wish to add an extra sum to cover the possibility of an unfortunate outcome to the project.

Profits

The main aim of the contractor's organisation is to make profits for the owners of the company and a percentage will be added for this purpose. However, the percentage added must be carefully considered given the prevailing market conditions. If a large sum is added, the profits made will be large if the contract is won, but the likelihood of winning the contract is also reduced.

Many civil engineering projects still adopt a measurement-type contract using a bill of quantities, and although the contractor may see the project costs in the above categories, s/he will often be required to submit the tender in the form of rates against bill items. The most obvious way for the contractor to distribute his/her costs in a CESMM3 BOQ would be as follows:

Direct costs   against the relevant BOQ item (excluding temporary works costs)

Temporary works   against general items (method-related charges)

costs + on-costs

Mark-up   against the adjustment item

It should, however, be realised that different contractors will have different policies on how their costs are to be recovered through the BOQ. Although the conditions of contract assume that the rates properly represent the contractor's actual costs/prices for each item, the contractor is not bound to comply with this assumption. Indeed, there are some circumstances in which it may be to the contractor's advantage to price rates unevenly. Two examples of situations where this might occur will now be described.

Front-end loading

Figure 9.2 shows the contractor's predicted income and expenditure on a contract, with the smooth line representing the costs s/he expects to incur (paying for labour, plant, etc.) and the stepped line representing the anticipated income (payments from monthly certificates). From this diagram, we can see that the contractor's income lags behind the costs for most of the contract time and it is only in the later stages that the contract has a positive cash-flow. This means that for most of the contract, the contractor must provide a varying level of finance which will involve additional costs, often in the form of interest on borrowings. If the contractor could earn income faster from the contract, this additional cost could be reduced or eliminated.

In Fig. 9.3, we see that this has been achieved. The contract may actually have a positive cash-flow throughout most of the construction period, and yet the overall income from the contract is the same as shown in Fig. 9.2. This can be achieved if the contractor identifies certain items that will be paid early in the contract and increases the rates of these items above their realistic cost. To keep the overall cost of the contract the same, items that will be paid towards the end of the contract must also be identified and their rates reduced. In this way, income is effectively moved from a later stage of the contract to an earlier stage. This is commonly known as 'front-end loading'.

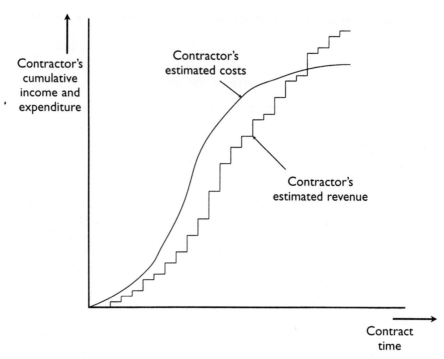

Fig. 9.2 Contractor's predicted income and expenditure

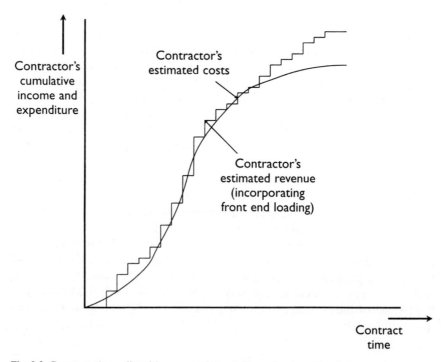

Fig. 9.3 Contractor's predicted income and expenditure: front-end loaded

Use of a mis-measured bill item

If during the pricing of a tender, the estimator identifies a substantial item that appears to have been mis-measured, s/he may increase the rate of that item in such a way that it will effectively make money for the contractor's organisation with no additional effort.

As an example, consider a contract on which the tender BOQ contains an item (item B) for 1 000 linear metres of kerbing. The estimator realises that there are actually 5 000 linear metres of kerbing on the job and that the designer has made a gross error. To take advantage of this, the estimator decides to make use of another item (item A) in the BOQ for excavation, whose tender and remeasured quantity is 3 000 m³. This is summarised in Table 9.4.

The realistic rates of these items are shown in Table 9.5, together with the total item costs in both the tender BOQ and when the quantities are remeasured.

By carefully loading the rate of item B and reducing the rate of item A, as shown in Table 9.6, we can see that the amount of money that shows up in the tender BOQ is the same as that using real cost rates. This means that the tender will still be competitive. However, when the items are remeasured, we see that the contractor has made an additional £18 000.

It is not known how often contractors employ these methods and it should be noted that they are not without risk. For example, if the Engineer ordered a variation that involved the addition of a large quantity of item A, say, in the above case, the contractor might well be expected to do that extra work at the rate that s/he knew would lose money. Having said that, the author has had first-hand experience of working on a contract where the rate for excavating unsuitable material was £0.01/m³!

Whether the contractor does or does not decide to adjust any rates – and such adjustments are probably the exception rather than the rule – s/he must first be able to make an estimate of the realistic cost of constructing the project. This will involve making decisions on the level of on-costs, on the percentage for company overheads and additions for risk and profit. The rest of this section will, however, concentrate on the methods used for calculating direct cost estimates; that is, establishing the direct cost of the contract work.

Establishing a direct cost rate for contract work

The main methods used by contractors to determine the actual cost of carrying out the work of the contract are use of a subcontractor's quotations, unit rate estimating and operational estimating, and these will now be described in detail.

Table 9.4   Bill of quantity items to be adjusted

Item	Quantity in tender BOQ	Remeasured quantity
A	3 000 m³	3 000 m³
B	1 000 lin metre	5 000 lin metre

Table 9.5   Bill of quantity items: realistic rates

Item	Real cost rate (£)	Item total tender BOQ (£)	Item total on remeasure (£)
A	3.00	9 000	9 000
B	8.50	8 500	42 500
	Totals	17 500	51 500

Table 9.6   Bill of quantity items: adjusted rates

Item	Adjusted rate (£)	Item total tender BOQ (£)	Item total on remeasure (£)
A	1.50	4 500	4 500
B	13.00	13 000	65 000
	Totals	17 500	69 500

Use of a subcontractor's quotations

The use of subcontractors to carry out parts of the contract work is now commonplace in the construction industry, and the main contractor will be responsible for the work of any subcontractors s/he employs. Having decided at tender stage which parts of the contract will be sublet, the main contractor will obtain prices for that work from the subcontractor selected, and these will often be in the form of rates against the relevant parts of the bill of quantities. The main contractor needs only to ensure that the rates s/he puts in the tender are compatible with the subcontractor's rates to be reasonably confident of avoiding financial risk.

Unit rate estimating

For the particular work to be costed, the estimator must first decide what resources will be used and at what level; for example, if excavating plant is required, what size should be used and how much is needed? Having established the resources, which will be mainly labour, plant and materials, further decisions must be made regarding the expected productivity of these resources on the work in question. In this approach, the decision will be in the form of a number of hours/minutes/seconds (say) the excavator will take to move 1 m³ of material. Contractors will hold information about the productivity of most resources in their files and the estimator will use these figures, adjusted as s/he thinks fit because of the specific site constraints, to develop a productivity to be adopted for this particular item of work. With known resources operating at defined levels of output, the amounts of material and time for which labour and plant will be needed to produce 1 unit of output can be determined. What we must still do is to provide a cost for the use of these resources, and we can then cost the work. Costing the materials element is relatively straightforward, as quotations may be used, and provided that any wastage has been allowed for, the estimated cost should be reliable. Labour and plant costs are not so easily determined and, as an example, the real cost to the contractor of employing labour, often called the 'all-in' cost, will need to include, as a minimum:

- basic wage

- allowances to cover guaranteed time

- sick pay

- national insurance

- allowances for holidays with pay.

Unless the contractor makes such allowances, s/he will not be covering the actual costs.

After calculations have been carried out for all-in labour and plant costs for the variety of different classes of labour and types of plant, adjusted for the area of the country in which the work will take place, the cost of producing 1 unit of the work can now be deduced. If this is multiplied by the number of units required on the project, the cost of that item can be predicted. As an example, consider the following:

Task: to determine a rate for 'Precast concrete kerbs straight or curved to radius exceeding 12 m', of which there are 280 linear metres (lin m) in the bill of quantities.

Method: the following resources will be needed:

- kerb-layer and labourer

- kerbs

- grade 7.5 concrete bed and backing

- fork-lift to off-load the kerbs from the lorry.

The estimator must then decide on output or usage rates, as follows:

- kerb-layer and labourer   0.2 h per lin m

- kerbs   1.05 lin m per lin m

- concrete   1.2 m^3 per m^3

- fork-lift   0.02 h per lin m.

Note that the rates for materials are given to recognise that there will be some wastage.

The concrete and kerbing costs come directly from quotations for supplying these materials to the site, and all-in rates for the labour and the rough-terrain fork-lift are calculated on the basis of anticipated local pay rates and hire rates respectively.

- kerb-layer and labourer   £21.30 per h

- kerbs   £3.15 per lin m

- concrete   £1.52 per lin m

- fork-lift   £14.36 per h.

Thus the overall cost of the resources per linear metre is:

- kerb-layer and labourer   £21.30 × 0.2   =   £4.26
- kerbs   £3.15 × 1.05   =   £3.31
- concrete   £1.52 × 1.2   =   £1.82
- fork-lift   £14.36 × 0.02   =   £0.29

                                      Total   =   £9.68

The total cost of the item is therefore 9.68 × 280 = £2710.40.

Operational estimating

In this approach to estimating, a set of resources is committed solely to a major activity and the total cost of using materials and keeping labour and plant on site for the duration of the activity is recovered fully from the rates applied to that single activity. Although we have specified that one activity only will be involved, this may be represented in the bill of quantities by a number of items. For example, the placing of concrete is one activity, but would typically appear as several items in any bill of quantities containing such work. Having decided on the resources required, the estimator must decide how long they will be needed for in order to complete the whole of the work and must then apply sensible materials rates and all-in plant and labour rates to devise the rate for each bill item. The following example should aid understanding:

Task: to determine a rate for 'Excavation for cuttings, material other than top-soil, rock or artificial hard material' of which there is 3420 m³ in the bill of quantities.

Method: although this bill item is only for 3420 m³, there are several other excavation items in the BOQ that will need to use a similar combination of resources, 12 155 m³ in all. For all of this work, the estimator decides that with a particular combination of resources, a production rate of 1600 m³ per week is possible. It will therefore take 7.6 weeks to complete the work. All of the excavation is expected to be used as fill and the excavation item will include for haulage to fill sites. The resources used will be:

- 1 No. 2.0 m³ excavator   (£25.45 per hour)

- 3 No. 10 m³ dump trucks   (£19.30 each, per hour)

- 1 No. banksman   (£5.78 per hour)

and the plant will be obtained from a plant hire company and will come with drivers included. Thus the overall cost of these resources can now be calculated.

Resource	No.	Rate (£/week)	Duration	Cost
Excavator	1	1 018	7.6	7 737
Dump truck	3	772	7.6	17 602
Banksman	1	231	7.6	1 756
			Total cost =	£ 27 095

The overall cost per cubic metre = 27 095/12 155 = £2.23, and this rate may be applied to all excavation items comprising the 12 155 m³.

Operational estimating is usually seen as the preferred method for significant bill items. The main reason for this is that the contractor knows definitely that the full cost of the resources used will be recovered through the bill of quantities. When using unit rate estimating, the various activities will require resources for certain disparate periods. This means that a separate check will need to be made that (say) for plant, any periods of inactivity during which the plant will be on site and costing the contractor money, but not used on productive work, are identified and allowed for somewhere in the estimate.

The final point to be made in this chapter concerns the need for the contractor to make the best use of the tender period in ensuring that the most accurate estimate and therefore the most well-informed tender are produced. The tender period is usually a time of intense activity for the contractor's estimating department and its personnel are likely to need to prioritise their work. Studies have shown that a large proportion of the total value of a contract will be contained within a relatively few bill items, and this has become known as the 80–20 rule. The rule says that in general, we can expect that 80% of the cost of the contract will be contained within 20% of the bill items, as shown in Fig. 9.4. If the estimators

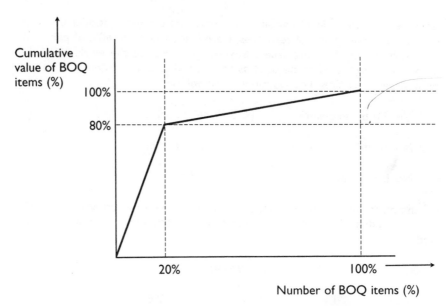

Fig. 9.4 The 80–20 rule

identify these most important items and concentrate their efforts here, using less accurate methods for the other items, this should be the most effective policy.

Recommendations for further reading

McCaffer, R. and Baldwin, A.N. (1991) *Estimating and Tendering for Civil Engineering Works*, 2nd edition. Granada, London. 330 pp.
A comprehensive text aimed specifically at the civil engineering side of the construction industry.

10 General management topics

Introduction

All of the information presented so far in this book has dealt with the way in which the construction industry operates, the techniques used or other information useful to an understanding of the industry. Much of this material is often presented under the heading of 'project management'. What have not been addressed are what might be called 'general management topics'. These concern themselves with the problems of running *any* business or organisation, and may be considered under the following headings:

- policy planning

- organisation theory

- human resource issues

- marketing and sales

- production

- finance.

Policy planning has already been mentioned in Chapter 5, 'Planning and control'. It involves the organisation identifying its overall objectives, usually in terms of products/services to be provided and market share to be sought, and developing policies which it hopes will allow it to achieve those objectives.

One of the issues that must be resolved by the company is how to arrange the various people who work in the company to achieve its objectives. This involves decisions about how the employees should be grouped, and how authority and accountability should be assigned and measured.

Human resource issues will address the problems of how to get the best staff to join the company, and how to retain them and encourage them to be productive. Words often used in this environment are selection, personnel, team-working and motivation. For many organisations, staff costs are their principal expense and efforts to gain the most benefit from this costly resource are easily justified.

If the products that the company provides cannot find a customer, or sufficient customers, the company will clearly suffer. One of the main functions of a marketing department within a company is to forecast future demand for products the company is planning to produce and to try to gauge the requirements would-be customers are looking for in such a product. In this way, some of the risk in launching a new product should be removed. The business of ensuring customers for the company's products, new and old, which will involve keeping in touch with existing customers and looking for new ones, is dealt with under the heading of 'sales'.

For many companies, provided there is a market for their products, their profitability will depend on the efficiency with which they are able to produce the goods, which will normally have to compete on price with other, similar goods. Ensuring that acceptable-quality materials are supplied, that these materials are effectively assembled or transformed into the final products and that the quality of the finished product is to standard is the job of the people involved in production. Many of the techniques adopted in this area will depend on the type of product and how it is produced, but the emphasis on keeping costs low and quality high will inevitably pervade.

Although other important motives for being in business may be cited, a prime reason will always be to make money for the owners. Money is needed at all stages of the company's operations and must be provided by sales, via shareholders or by other forms of borrowing to allow the company to continue to operate. This aspect of business is presided over by the company's accountants, who will produce regular statements indicating the financial position and predicting the profits that have been made on the capital employed. It is, however, not just the accountants who must consider the company's finances; all employees who have authority to make spending decisions must do so with cost in mind.

The study of management as a separate subject is relatively new, with the bulk of the development happening in the twentieth century. Despite this, most university libraries will contain several rows of books devoted to the above topics. There is thus much to learn, but in this book, which concentrates on civil engineering, an introduction is all that can be offered. To give a little more insight, two topics will be dealt with at greater length. These are organisation theory and motivation.

Organisation theory

Before the First World War, most companies in the UK were family-owned and their structure was typically very simple with obvious and clear lines of command and responsibility. The owner was the main source of authority in the company and the workforce and any clerical staff would have limited opportunities to make decisions. As further developments gave rise to the need for a greater proportion of specialists, e.g. accountants, sales, personnel, production controllers, etc., it became increasingly necessary to consider the way in which the company was organised. To ensure that the work of the organisation was carried out efficiently, a hierarchical structure developed with managers at the various levels in the company operating with delegated authority from other managers higher up in the structure. As firms became larger and more complex, the importance of getting a good organisational structure, which would allow proper control and also allow good communication, co-ordination and use of resources, meant that the study of these matters flourished. Research showed that the choice of a particular structure for a company would typically depend on several factors, including the company's size, the type of work it carried out and the way in which its markets were changing.

Organisation charts

The traditional way of showing a company's organisational structure is by hierarchical charts, which identify the managers and show how they relate to each other within the company. Figure 10.1 shows the higher levels of the organisation chart for a major UK contractor, with interests in civil engineering construction, housing and mining in the UK, and some overseas interests. These are represented in the organisation as major divisions, and at the head office there are also central services, including engineering services, central accounts and central legal services. All of these parts of the organisation will involve considerable complexity with several lower levels in their hierarchies. One further level is shown under the UK civil construction division, where separate plant, purchasing, estimating, contracts management, services and administration departments are in evidence. Here, the manager with complete responsibility for this area of the company's business clearly reports directly to the Chief Executive and has managers in each of the areas just mentioned reporting directly to him/her. Because the UK civil construction sector is a major part of the contractor's business, there are three regional managers in each of the three regional offices (*see* Fig. 10.2) who report to the UK civil construction manager, and sizeable hierarchies exist at each of these offices. For all major sites that are operated through the regional offices, there will also be a substantial management structure. Figure 10.3 is an example of a typical contractor's site structure.

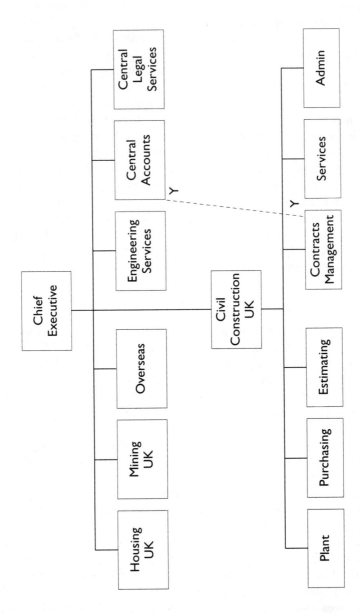

Fig. 10.1 Contractor's high-level organisation

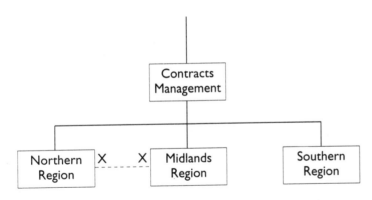

Fig. 10.2 Organisation of regional contracts management

These diagrams help us to understand how the company or site is organised and give us some understanding about who must report to whom and how the work of the organisation is controlled through its managers. In Fig. 10.1, we have already seen that the manager of Civil Construction UK is accountable for all civil engineering construction the contractor carries out in the UK, and this is clearly an enormous responsibility. To allow him/her to have some control over this massive workload, the management structure that has just been outlined means that parts of the work are delegated to managers in the next level in the hierarchy; here, the six managers in plant, purchasing, etc. These managers in turn delegate, and the formal structure effectively defines the line of command and areas of responsibility. For the system to work well, each manager needs to be clear about the limits of his/her specific responsibility and should be given the authority necessary to allow those tasks to be carried out efficiently. Responsibility without authority makes the manager responsible for something s/he does not have the power to achieve. For example, when particular resources must be used to complete a part of the work and the manager does not have authority to hire or buy or obtain those resources from internal plant holdings, s/he is prevented from fulfilling the role.

The number of immediate subordinates a manager has reporting to him/her is known as the *span of control*. For the UK civil manager, the span of control is clearly 6. When this number is too large, control will not be properly exercised, and for fairly stable, traditional industries like construction, a figure of 5–8 is often seen as acceptable. In more unstable industries, where markets and products are rapidly changing, the recommended span of control would be lower.

Although the formal structure identifies the main lines of communication and command, and for a substantial organisation seems extremely complicated, there will typically be much that is not shown. The informal structure of the organisation will involve many more communication links and a great deal more information about how the various sections interact

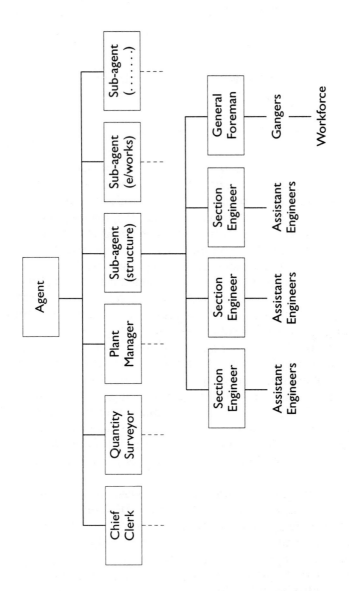

Fig. 10.3 Contractor's site organisation

and work together. It is this additional knowledge that people working in the organisation learn, and for members of staff who have worked in several sections of the company, the insight they develop about the real workings of the company and its problems can be very useful. Some of the additional lines of communication will be between managers at the same level in the hierarchy (X–X in Fig. 10.2), who must co-operate to ensure sensible use of any shared resources. There will also need to be discussions between managers in different parts of the organisation (Y–Y in Fig. 10.1), and these lines of communication should be known and accepted by any manager effectively bypassed in the process. These links are often known as *lateral links* and *functional links*, respectively, and will be essential if the company is to operate properly.

Organisational structures for project management

The organisation charts shown so far identify what are known as functional managers; that is, managers who look after a specific function of the organisation – estimating and plant management, for example. Much civil engineering, however, is project orientated, and the way in which projects are handled within the company structure will have a direct effect on the efficiency of the company as a whole. The literature recognises two distinct ways of incorporating projects and these are shown for a design organisation in the highways field. Figure 10.4 shows project managers with direct authority over the various specialists needed to produce the complete design service, and this arrangement has both advantages and disadvantages, as shown below:

Advantages:

1. With direct authority over all the people needed, the project manager should not be hindered by lack of resources.

2. A definite team is established and this can have a positive effect on motivation.

Disadvantages:

1. The specialists are distributed throughout the various project teams, and this is unlikely to lead to the development and maintenance of expertise.

2. Resources may not be used efficiently as the project manager holds onto people who will be needed soon, but are not really needed at the moment.

3. There are poor career prospects for individuals in the team, who must work under a succession of project managers.

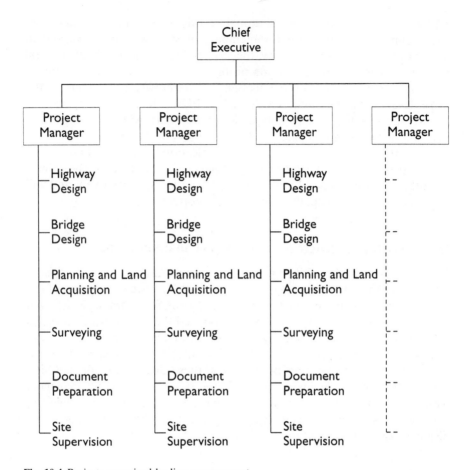

Fig. 10.4 Projects organised by line management

Figure 10.5 shows what is often described as a 'matrix management' arrangement, in which functional departments exist, but project managers are able to obtain the services of the various specialists as and when they are needed by the project. Again there are advantages and disadvantages:

Advantages:

1. This system is flexible, making it easier to respond to unstable markets.

2. Greater use of personnel should be made, with specialists linked to a project only when needed.

3. Because there is a functional grouping for the various specialists, the development of expertise should be enhanced.

4. Career progression is possible via either functional or project management directions.

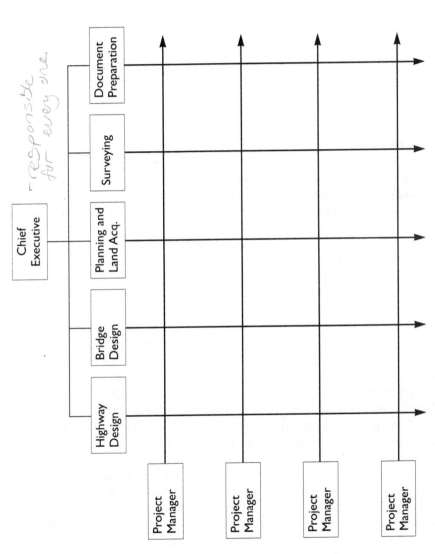

responsible for every one

Fig. 10.5 Projects organised by matrix management

Disadvantages:

- Individuals within a functional department have divided loyalties, increasing the possibility of conflict.

- Accountability may be lost as individuals try to satisfy both project and functional managers.

A final note on this subject: students should realise the value of gaining a good understanding of the organisational structure of the company to which they belong. If they are able to move about in the company and see a variety of different sections and departments, they will begin to understand in detail the work done and the problems faced by managers at different levels. This understanding should be evident in any interview they may have for a management position.

Motivation

The desire to improve the productivity of individuals working within an organisation in order to reduce costs and thereby increase competitiveness and profits is one of the main objectives of any management system. Research on this topic has tried to identify the primary factors that motivate people to work harder; not simply for the sake of interest, but also in order to be able to influence workers' behaviour in a direction that will be beneficial to the organisation. From the studies undertaken, theories of motivation have been proposed and some of these will be described in this section.

Surely one of the main attributes that will be looked for in any prospective candidate for promotion will be that the person is well motivated. This often shows itself in a desire to get on with the work and perhaps also a willingness to continue to work beyond the normal working day. Such candidates, provided they have the skills and knowledge needed, will soon find themselves in a management position. By this is meant a position in which the person is responsible not only for his/her own work but also for the work of others. In this changed position, the new manager can only shine if the output of the whole section is high, and this clearly means that personal motivation will seldom be enough. Somehow the manager must convince the people in his/her section to work with the same interest and vigour as s/he exhibits, and only if this can be achieved is further promotion likely. We can therefore easily see that as well as sound corporate motives for wanting to influence the behaviour of others, there are also personal motives that may be even stronger.

All managers must confront the problem of motivating staff, and will invariably have strong views on the topic, ranging from a belief that the workforce must 'obey' the management to a belief that management must 'encourage' the workforce to be productive. These extremes were further defined by Douglas M. McGregor, who used the terms 'Theory X' and 'Theory Y' to label them. Theory X is defined by the following propositions:

1. It is management's job to arrange the money, materials, equipment and people in the oganisation to achieve its objectives.

2. People must be directed, motivated and controlled to fit the needs of the organisation.

3. Unless people are treated in this way, they will not respond to organisational needs. They must therefore be rewarded or punished as necessary.

These views embody an opinion of the average employee as:

4. lazy by nature and preferring to work as little as possible;

5. lacking ambition and not wanting responsibility;

6. self-centred and not interested in the organisation's aims;

7. resistant to change and not very bright.

The alternative view, Theory Y, can be summed up as follows:

1. It is management's job to arrange the money, materials, equipment and people in the organisation to achieve its objectives.

2. People can respond to organisational needs and will do so if management arranges the conditions to allow this; that is, if they are permitted to achieve their own goals, which are in line with the organisation's goals, by directing their own efforts.

3. People exhibit potential for development and willingness to shoulder responsibility. They only appear passive as a result of their previous bad work experiences.

The corresponding opinion of the average employee is thus of someone who:

4. is not naturally lazy, and is willing to work hard if treated well;

5. accepts responsibility and is interested in the decision-making process;

6. is willing to work towards organisational goals.

Although labelled Theory X and Theory Y, these views are not actually theories. They are really just extreme examples of the attitudes managers might have towards the workforce. While Theory X relies heavily on external control, Theory Y depends on the workforce exerting considerable self-control. McGregor saw the distinction as the difference between treating people as children and treating them as mature adults. As already mentioned, however, theories of motivation in the workplace do exist, and three of these will now be described in some detail.

Maslow's hierarchy of needs

By questioning people about the peak experiences in their lives and analysing the results, A.H. Maslow recognised a pattern that led him to develop a theory attempting to explain how individuals are motivated. The theory relies on a concept that is now called the 'hierarchy of needs'.

The theory says that we humans have a number of needs, and that in certain circumstances the desire to satisfy a particular need will motivate us into action. The needs themselves and the way in which they operate on the individual are shown diagrammatically in Fig. 10.6. Maslow says that the need to satisfy our hunger and thirst is, as he puts it, the most prepotent. By this he means that for a person with serious concerns about how they will feed themselves, that person's actions will be motivated solely by attempts to satisfy this most basic need. Only when the physiological need has been acceptably satisfied will the next higher-level need, the need for security, become a motivator, and only when the person's concern about the security of themselves and their family is overcome will the next higher-level needs begin to operate.

The hierarchy continues with needs relating to self-esteem and self-actualisation at the top end. This highest need is said to be satisfied when the individual 'becomes everything that s/he is capable of becoming', and for the highest-level needs, it is argued that the gratification that comes from satisfaction leads to a need to repeat the experience.

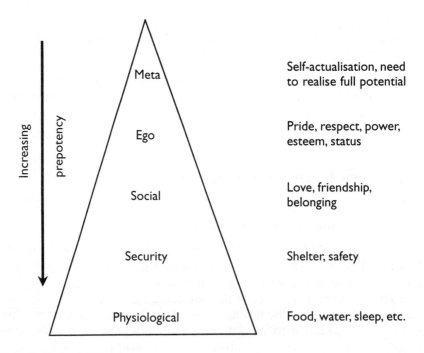

Fig. 10.6 Maslow's hierarchy of needs

For the manager who wishes to motivate his/her staff, the theory provides some assistance. For example, it will clearly be unhelpful to try to motivate an employee who is unable to afford acceptable accommodation for his/her family by suggestions about the prestige and status of their current job, if no more money is on offer.

The theory is by no means universally accepted, and even Maslow felt that in some circumstances the order of the needs may vary for certain individuals. He also argued that for a person who has lived life at a low level, say having been unemployed for a long time, that person might be satisfied with sufficient food and adequate shelter for the rest of his/her life, and not be concerned with higher-level needs.

Frederick Herzberg

Herzberg carried out interviews on a large sample of engineers and accountants representing a cross-section of Pittsburgh industry. During the interviews, respondents were 'asked about events they had experienced at work which either had resulted in a marked improvement in their job satisfaction or had led to a marked reduction in job satisfaction'. By probing the interviewees for the reasons they had felt as they did, Herzberg concluded that the factors that affected motivation could be grouped under two separate headings. He called these 'motivators' and 'hygiene factors', as shown in Fig. 10.7.

Motivators	**Hygiene factors**
Achievement	Company policy
Recognition	Supervision
The work itself	Salary
Responsibility	Interpersonal relations
Advancement	Working conditions

Fig. 10.7 Herzberg's motivators and hygiene factors

The theory says that the motivators will tend to lead to increased job satisfaction when positive, but will not cause particular dissatisfaction when negative. Thus when an individual's achievement is recognised, this will lead to satisfaction, which will result in improved performance, but lack of recognition does not generate feelings of dissatisfaction and poor performance.

On the other hand, the hygiene factors are said to lead to dissatisfaction when negative, but do not promote particular satisfaction when positive. That is, they can only have a negative effect if perceived as pitched at an unacceptable level. As salary is a hygiene factor, this means that when people perceive levels of pay as too low, it makes them feel dissatisfied, but when they perceive them as too high it does not lead to particular satisfaction. Clearly satisfaction is seen as linked to good performance and dissatisfaction to poor performance. The use of the word 'hygiene' here comes from the medical use of the term. A hygienic environment is safe and further efforts to improve the environment will have no positive effect. The hygiene factors operate in a similar way on worker performance. Provided that they are seen as being at an acceptable level, there is no value in improving them.

The theory thus suggests that management should concentrate on the motivators to improve satisfaction and performance while ensuring that the hygiene factors are at an acceptable level to avoid dissatisfaction.

The theory of complex man

The two previous models of behaviour make statements about the factors that are seen to motivate generally, but Schein's theory attempts to deal with the variability of human nature and is illustrated by the following assumptions:

1. People are complex and highly variable. They have many motives which are arranged in some sort of hierarchy, but their order in the hierarchy is not fixed and may vary.

2. People's needs are satisfied in different ways and not all will look for satisfaction in the workplace.

3. Different management strategies will create a positive response in some while not in others. That is, there is no universal strategy that will work for all people at all times.

The above statements, although easy to accept as true, do not appear to offer much in the way of guidance to the practising manager. However, if the variability of individuals is considered to be a major factor, it is easy to see how acceptance of this leads on to a recommendation that managers need to be trained in interpersonal skills, so that they can assess individuals and react accordingly. Training, it is argued from this theory, should thus be aimed at developing two major qualities:

• the ability to appreciate personalities and motivations of subordinates, using case studies and role play;

• the ability to adapt to different problems, developed by exposure to a variety of such problems.

Many other views and theories have been expressed about the way in which people can be motivated to become more productive, and the area is one of continuing interest to researchers and professionals. Few would suggest that any one approach will ever adequately address the complexity of this area, but it is believed that exposure to this kind of material is helpful for the new manager to develop his/her own personal attitudes.

Recommendations for further reading

Lawrence, P.A. and Lee, R.A. (1989) *Insight into Management*, 2nd edition. Oxford University Press, Oxford. 231 pp.
Most general management books will address motivation and organisation theory, but this book is aimed specifically at science and engineering students and is more relevant as a result.
Vroom, V.H. and Deci, E.L. (1975) *Management and Motivation*. Penguin Education, Harmondsworth. 399 pp.
It does not matter that this book is rather dated, because it contains extracts from the writings of the main contributors to the study of motivation.

Index